AF713011

THE BATTLE OVER THE BUTTERFLIES OF THE SOUL

Camillo Golgi, Santiago Ramón y Cajal and The Birth of Neuroscience

Wallace B. Mendelson

Pythagoras Press

Copyright © 2023 by Wallace B. Mendelson

All Rights Reserved.
No part of this book may be reproduced in any form or by any electronic or mechanical means including information storage and retrieval systems, without permission in writing from the author. The only exception is by a reviewer, who may quote short excerpts in a review.

Printed in the United States of America.

First printing: January 2023

About the Author

Wallace Mendelson, MD is Professor of Psychiatry and Clinical Pharmacology (ret.) at the University of Chicago. He is a Distinguished Fellow of the American Psychiatric Association, and a member of the American Academy of Neuropsychopharmacology. He has been director of the Section on Sleep Studies at the National Institute of Mental Health, the Sleep Disorders Center at the Cleveland Clinic Foundation, and the Sleep Research Laboratory at the University of Chicago. He is the author of fourteen books and numerous professional papers. He is currently a Fellow of the Faculty of History and Philosophy of Medicine and Pharmacy at the Worshipful Society of Apothecaries, London. Among his honors have been the Academic Achievement Award from the American Sleep Disorders Association in 1999 and a special award for excellence in sleep and psychiatry from the National Sleep Foundation in 2010. More information about Dr. Mendelson's work is available in Wikipedia at https://en.wikipedia.org/wiki/Wallace_B._Mendelson and on his website at http://zhibit.org/wallacemendelson.

CONTENTS

INTRODUCTION: A CONFRONTATION IN STOCKHOLM

In December 1906, two eminent scientists, the Italian Camillo Golgi and the Spaniard Santiago Ramón y Cajal, arrived in Stockholm to jointly receive the Nobel Prize in Physiology or Medicine. Golgi had discovered a breakthrough method of visualizing neurons under the microscope in 1873. He also believed that the nervous system was different from the rest of the body: instead of being composed of a multitude of individual cells, he thought it was one vast 'reticulum' of cells which were anatomically fused together, which made 'unity of action' possible. Cajal, a decade younger, had started with Golgi's method, improved on it, and concluded that neurons were contiguous, but not actually continuous, and communicated across gaps between them. His view, known as the 'neuron doctrine', specified that individual neurons were the basic anatomic and functional building blocks of the nervous system, and ultimately became the foundation of modern neuroscience. Though the two had been vigorous rivals for two decades, they had never actually met face to face, and never would again.

Cajal, who had actively sought the prize, was met at the Stockholm train station on December 6 by Gustaf Retzius, an eminent Swedish histologist, and other luminaries. Golgi, who had always considered the prize a 'fantasy' he had not expected to receive, tried to keep his arrival secret, but word had leaked out. On December 8, following a difficult sea crossing in stormy weather, he, his wife, and an assistant got off the train on the last leg of their journey. A distinguished delegation was waiting for him, including Cajal, who had apparently hoped that they might meet and talk. To everyone's surprise, Golgi, who was ill and whose wife who was in pain after a fall during the trip, greeted them curtly and rushed off to the hotel, wanting to work on his forthcoming lecture. He was anxious about the talk, which would be conducted in French, and spent the next two days working on it whenever he could escape the deluge of social activities.

On December 10, in the grand salon of the Royal Academy of Music, King Oscar II of Sweden presented Golgi and Cajal with their medals. In the next two days, the laureates gave lectures which traditionally were an opportunity to summarize their work. Golgi spoke first, on December 11, and to everyone's surprise gave a talk entirely devoted to attacking the neuron doctrine. He had been minimally involved in research in the last few years, devoting himself to administration and politics, and he seemed unfamiliar with recent findings. He described the acceptance of Cajal's views in different ways, at one point referring to 'when the neuron doctrine made, by almost unanimous approval, its triumphant entrance on the scientific scene...', and also saying that 'the majority of physiologists, anatomists and pathologists still support the neuron doctrine', yet at another asserting surprisingly—and incorrectly--that it 'is generally recognized to be going out of favor' (Golgi, 1906). Indeed, he was critical of even using the word 'neuron'. In his talk it was almost as if the voluminous work by Cajal and others in the last two decades had had no impact on his thinking, as he re-stated his own conclusions, first reached in the years around 1873.

Cajal remained silent until the next day, when he gave a lecture going through the studies which step by step had led him to his viewpoint. He concluded, referring to Golgi, this way: 'We mourn this scientist who, in the last years of a life so well filled, suffered the injustice of seeing a phalanx of young experimenters treat his most elegant and original discoveries as errors' (Ramón y Cajal, 1906). It was not Golgi's last years—he lived until 1926—but it was true that Cajal was prescient: the neuron doctrine continued to be accepted and became the dominant view of neuroscience at the time, while reticular theory, supported by the anatomist who had made It possible to view nervous system tissue under a microscope, was left by the wayside.

What led up to this remarkable confrontation? How had two eminent scientists gotten so caught up in their struggle that they retreated into fantasy, the one saying that the opposing view (now overwhelmingly dominant) was going out of favor, while the other incorrectly pictured his rival as being in his last years? As a psychiatrist and scientist, I became intrigued with these questions, and with a larger one: what led Golgi to cling to his old beliefs and refuse to recognize a growing and overwhelming body of evidence to the contrary? This book reflects my efforts to find answers.

In writing this I have relied on the Nobel biographies and lectures of the two protagonists, a variety of their writings, as well as two books. *Golgi: A Biography of the Founder of Modern Neuroscience* was written by Paolo Mazzarello, a physician, neuroscientist, and historian. *The Brain in Search of Itself: Ramón y Cajal and the Story of the Neuron* was authored by the journalist and writer Benjamin Ehrlich. Mine differs in several ways. In addition to thoughts arising from my background as a psychiatrist, I was struck by the importance of seeing this piece of history as taking place at the confluence of developments in three fields: microscopy, photography, and methods of staining tissue. I have been particularly interested in the latter since writing a book (*Molecules, Madness, and Malaria*; 2020) which describes how many medicines for psychiatric illness and infectious disease were derived from the growing sophistication of synthetic organic chemistry, which began in the nineteenth century fabric dye industry. Just as advances in photography led to the availability of the silver nitrate used by Golgi to visualize neurons, this evolution based on synthetic fabric dyes led to stains for microscopy, such as methylene blue, used as supporting evidence by Cajal to refute Golgi's reticular theory. Although most of these influences (though perhaps not this aspect of synthetic organic chemistry) are mentioned in the previous works, I have dwelt on them at some length, and tried to make the case that Golgi and Cajal's work was made possible by the moment in time at which these three fields came together, intersecting as well with two particular personalities. I have also summarized how their ideas have fared in the century since the Nobel Prize.

By emphasizing the historical context, I have also come across a number of other vivid characters whose stories should be told. Who would have guessed, for instance, that one of the first men to emphasize that individual cells were the basic building blocks of life was first an unsuccessful lawyer who shot himself, somehow recovered, and went on to a career in science? Or that one of the first to argue that neurons were in fact not fused together was a graduate student who reported this in his dissertation, then a few days later abandoned science, leaving for an adventure as an Arctic explorer, and later won the Nobel Peace Prize?

I should also mention that in presenting 'my take' on the events leading up to Golgi and Cajal's conflict, how it was conducted, and its outcome, I have written in a manner designed to be accessible to the general reader who does not necessarily have a technical background. The reasoning is that these stories are too good to be confined to the experts. Of course, the hope is that at the same time it will also be useful to students and practitioners of the field who would be interested in learning more of its origins, though expressed in relatively non-technical terms. Those who would like to learn more about the lives of later scientists, who were affected by World War II and the post-war blossoming of neuroscience, may be interested in reading *Trial by Fire* (Mendelson, 2021).

Ultimately, I keep coming back to Golgi's character, and how he could maintain a fixed belief for decades after it was so well refuted. This can be seen in many different ways. On the one hand we now live in an age in which it feels as if many people persist in seemingly false beliefs in the political and social realm, apparently unaffected by what seem to be clear facts. On the other hand, we revere figures like Cervantes' Don Quixote, persistently driven by an ideal at odds with the real world (and re-told in modern terms in Salmon Rushdie's *Quichotte*). It seems to me that Golgi was enthralled by an attractive conception of the nervous system to which he could not let go, and he probably does not fully fit into either of these types of situations. As we will see, his response to criticism may also have been affected by the timing of the challenge, and qualities of his personality. I have tried in this book to picture what he was like, based on the information that is available. But first, in the next chapter, we will look at the context in which Golgi and Cajal interacted.

REFERENCES for the INTRODUCTION

Bibliographic note: In this and all subsequent reference sections, there are occasionally situations in which there are citations of web pages for which the year of creation is not clear. In those cases, the year listed is that of accession.

Ehrlich, B.: *The Brain in Search of Itself: Santiago Ramón y Cajal and the Story of the Neuron*. Farrar, Straus and Giroux, New York, 2022.

Golgi, C.: The neuron doctrine – theory and facts. Nobel Lecture December 11, 1906.
https://www.nobelprize.org/uploads/2018/06/golgi-lecture.pdf

Mazzarello, P.: *Golgi: A Biography of the Founder of Modern Neuroscience*. Oxford University Press, 2010.

Mendelson, W.B.: *Molecules, Madness, and Malaria: How Victorian Fabric Dyes Evolved into Modern Medicines for Mental Illness and Infectious Disease*. Pythagoras Press, 2020.
https://www.amazon.com/dp/B088QPG14K

Mendelson, W.B.: *Trial by fire: World War II and the Founders of Modern Neuroscience and Psychopharmacology*. Pythagoras Press, 2021.
https://www.amazon.com/dp/B08X1MRVXJ

Ramón y Cajal, S.: The structure and connexions of neurons. Nobel Lecture December 12, 1906.
https://www.nobelprize.org/uploads/2018/06/cajal-lecture.pdf

Rushdie, S.: *Quichotte: a novel*. Random House, 2019.

CHAPTER ONE: BACKGROUND

As we described in the Introduction, Golgi and Cajal's discoveries were made at a time of confluence of developments in three fields: microscopy, histological stains, and photography. Let's begin by looking at each of these in turn.

Microscopy

The development of light microscopes

The beginnings of using optical lenses for magnification are unclear, but may have begun as far back as the eight century BC. In 1850, the English archeologist Austen Henry Layard (1817-1894), when excavating the Assyrian palace of Nimrud (now in Iraq), came across a round piece of ground and polished rock crystal approximately 1.5 cm in diameter. Although it had crude magnifying properties, it is not known if it was used for this purpose (British Museum, 2022A). Although some have argued that a magnifying glass might have aided the Assyrians in writing small script, or in astronomy, there is nothing in their texts describing such uses.

Figure 1-1: *The 'Nimrud lens', thought to have been made around 750-710 BC, has both plane and convex surfaces, and could potentially work as the equivalent of a 3X magnifying glass, with a focal length of about 12 cm. It was apparently ground on a lapidary wheel, and also has 12 cavities which may have contained a liquid, possibly naphtha. It has been thought by some to be able to focus sunlight to some degree. Whether it was used for any of these purposes is uncertain, and it is possible that its only function was decorative.*

In one of the first descriptions of magnification, the Roman philosopher Seneca (4 BC- 65 AD) noted that looking through a round water-filled bowl would make writing appear larger. Interestingly the Emperor Nero (37-68 AD) would watch the gladiators through green emeralds, but it is likely that this was not for magnification but rather an early—and expensive—form of sunglasses. During the Middle Ages, the Arab astronomer Alhazen (965-1039 AD) wrote extensively of optics and lenses, and speculated that oval lenses could make objects appear larger, apparently without ever building one. It was not until 1240 AD that his writings were translated into Latin, and read in European monasteries. Soon monks were making 'reading stones', plano-convex quartz and later glass lenses placed on a page to magnify its contents. By the end of the 13th century, they were largely replaced by eyeglasses. First produced by craftsmen in Venice or Pisa, they consisted of lenses mounted in wood or leather, placed precariously on the nose; by the seventeenth century, Spanish craftsmen developed frames complete with temple pieces.

The development of the first compound microscope (one with multiple lenses) is usually credited to Zacharias Janssen, a Dutch spectacle maker whose creativity laten extended to counterfeiting coins, and his father Hans, in the period of 1590-1608. The device was described some years later as having been a tube one-to-two inches in diameter, mounted on a tripod, with lenses, one convex and one concave, at its two ends. A surviving model from that period was handheld, comprised of three tubes, two of which could move up and down for focusing, and capable of magnifying objects ten times.

In 1665, the English physicist and architect Robert Hooke (1635-1703) published a book entitled *Micrographia* ('small drawings'), based upon his studies using a microscope with a biconvex lens at the end, in combination with lenses in the eyepiece tube. Though more powerful, it suffered from chromatic aberration (inability to focus all colors at a single spot) and spherical aberration (blurriness at the edges of an image due to differences in the focal point of light passing through the center and the edges of the lens). In order to limit these distortions, he devised a diaphragm to limit light from the periphery, resulting in sharper, but darker, images, even when illuminated by an oil lamp.

Hooke presented sketches of a variety of objects including sections of cork, which were comprised of many small box-like chambers. He referred to these as cells, as they seemed similar to the abodes of monks. Although this was dead plant tissue, it marked an early step in what came to be known as 'cell theory', the notion that cells were the basic building blocks of plants and animals alike.

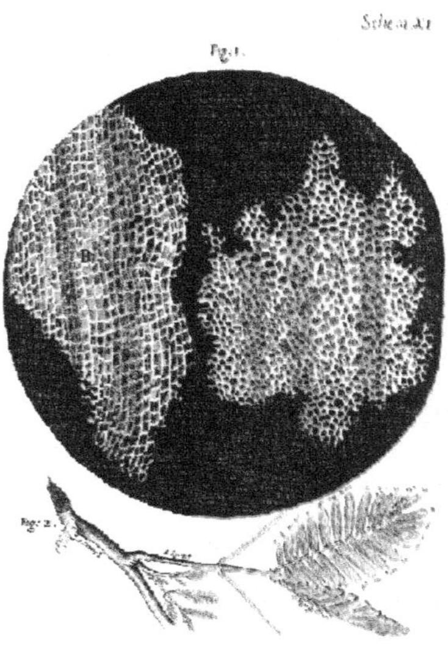

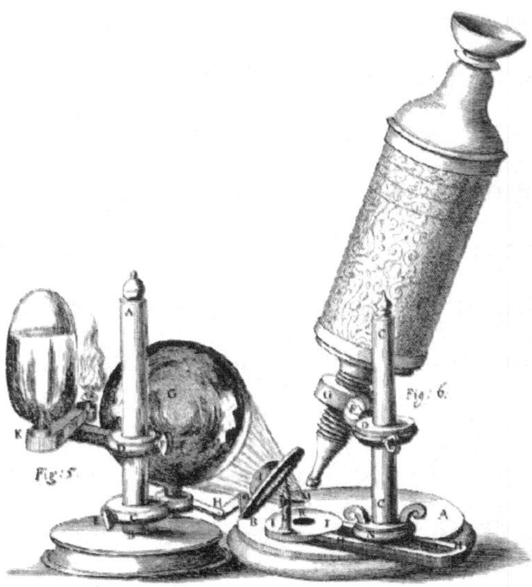

Figure 1-2: *Robert Hooke's microscope, as seen in* Micrographia, *1665. On the upper left is a drawing of his microscopic view of cork. Hooke published illustrations of a variety of objects, including snowflakes, tiny fossils, moss, the eye of a drone-fly, frozen urine, and an ant. The latter was particularly difficult to visualize, as it would not hold still, leading the frustrated microscopist to sedate it with brandy. The book concludes with illustrations of the sky which he visualized with his telescope (British Library, 2022b).*

In the 1670s, Antony van Leeuwenhoek (1632-1723), a Dutch draper and self-taught scientist, began to describe microbial life as seen through a one-lens microscope of his own design which could magnify an object up to 270 times. In addition to organisms in pond water (*animalcula*, or 'tiny animals'), he described red blood cells and their movement inside capillaries, muscle fibers, and spermatozoa. His many letters sent to the Royal Society of London are usually considered to mark the beginning of organized scientific microscopy.

The first steps in effectively dealing with the problem of chromatic aberration were taken in about 1729 by Chester Moore Hall (1703-1771), an English lawyer and inventor, who combined two lenses made of differing materials with different light-dispersing properties. He first applied this principle to telescopes, but by the 1820s microscopes with achromatic lenses, which refracted light of different frequencies into the same focal plane, were widely available, and this along with improvements in tissue preparation led to the blossoming of the life sciences. One example was the creation of cell theory, which, too, provides background for the story of Golgi and Cajal.

Microscopy and the beginnings of cell theory

In 1838, the microscopic studies of Matthias Jakob Schleiden (1804-1881) led him to propose a new way of understanding the role of cells, the first step in what later came to be known as 'cell theory'. Schleiden's career had begun with a law degree from the University of Heidelberg, and a practice in Hamburg. It turned out to be very unsuccessful; he became depressed and ultimately shot himself. Happily, he recovered and wisely realized that a change in profession was in order. After obtaining degrees in medicine and philosophy, he got a faculty position in botany at the University of Jena. During this time, he built on the work of the English botanist Robert Brown (as in 'Brownian movement'), who had described the nuclei in plant cells, to declare that cells comprised the basic structural units of plants, and that any given plant was derived from a single embryonic cell. This represented a radical departure from the then-prevailing notion of spontaneous generation, which proposed that the basic units of life were fibers or similar materials in dead or dying matter which somehow mysteriously came to life.

After 12 years in Jena, Schleiden came to believe that the stresses of academe were contributing to his anxiety and depression, and left university life to travel. In October 1838, he met Theodor Schwann (1810-1882), a young biologist at the University of Berlin, who at the time had been studying, and disproving, the notion of spontaneous generation, and who described the cells which surround peripheral nerves ('Schwann cells') and also coined the term 'metabolism'. Schleiden presented his cell theory to Schwann during a dinner;

afterwards Schwann showed him microscopic views of the spinal cords of fish, which seemed to confirm Schleiden's observations in plants. Schwann then proceeded to expand the notion to animal tissue, and with perhaps more clarity and detail proposed in 1839 that all living matter was comprised of cells, which are the basic building blocks of life. Some time later, in 1863, the German pathologist Rudolph Virchow (1821-1902), phrased his overall conception this way: 'The body is a state in which every cell is a citizen' (Virchow, 1863). In passing, it should be mentioned that Virchow, like Golgi (Chapter Two) led a life of both active scientific research and politics. He had participated in the revolution of 1848, and was a leader of the Progressive Party in the Diet, or Parliament, of Prussia, and an enemy of Otto von Bismarck. On one notable occasion, Bismarck challenged him to a duel, but fortunately for the history of science, Virchow demurred.

In 1855, Virchow had elaborated on an additional stipulation, that cells are created solely from pre-existing cells ('*Omanis cellular e cellular*'). Arguing that diseased tissues are not derived from organs in general, but rather from their cells, he gave credence to the process of studying microscopic pathology. The Swiss histologist Albert Kölliker (who as we will see in Chapter Three, played an important part in the acceptance of Cajal's findings) examined embryonic tissue in the light of cell theory, and added the notion that spermatozoa could be considered a type of cell, and cell theory, as originally formulated, was complete.

In modern times cell theory has been expanded to recognize that DNA is responsible for passing information across generations of cells, and that an exception has to be made for viruses, which are not necessarily cells, but hold genetic material and can generate new viruses inside of host cells. Nonetheless the basic proposition that organisms are comprised of cells was in place by the 1860s. One challenge facing histologists from that point on, however, was that what little they could see of cells in the brain made them look very different from those in other tissues; the nerve cells and their processes were so densely packed that it was hard to visually separate them. This was the situation facing Camillo Golgi when he began his landmark studies in the 1870s (Chapter Two).

Histological Stains

Another dilemma facing early microscopists was that most tissues appeared largely transparent and colorless, making it difficult to identify structures. It was necessary, then, to add dyes, or stains, which attached to different parts of tissue, coloring them and making it possible to visualize details. The first histological stains were natural substances (for instance, Leeuwenhoek, perhaps influenced by his Dutch heritage, had used a dye from the saffron crocus bulb). As time went on, synthetic compounds originating in the fabric dye industry appeared, as did inorganic metallic compounds derived from advancements in photography. In more modern times, well beyond the period we are describing, labeled antibodies have been used as well, but are not discussed here. Let's look at these earlier approaches.

Natural substances:

In 1858, Joseph von Gerlach, an anatomist at the University of Erlangen, Germany, asked an apothecary if there might be a substance which would make it easier to visualize blood vessels. He was given carmine, derived from the powdered bodies of female cochineal insects which live on prickly pear cacti in areas including Central and South America. It had been employed as a red fabric dye and a paint pigment since ancient times, and is still widely used to color candy, yogurt and fruit juices. It also appears in cosmetics such as eye shadow and lipstick. Insect-derived carmine had also been used by the English physician and botanist John Hill (1716-1775) as a microscopic stain for looking at wood as far back as 1770, but Gerlach was the first to apply it to animal tissue. As we will see in Chapter Two, Gerlach found carmine effective for seeing cells in many tissues, but at first his results in the nervous system were disappointing. One day he accidentally left a tissue sample immersed in a dilute carmine solution overnight, and found that with this lower quantity of carmine he could now visualize individual cells. He could also to some degree see their projections (dendrites and axons), but was unable to define their attachment to cells or their endings, leading him to conclude that they formed a vast interconnected network (Chapter Two). Carmine was one of the early stains Golgi used, before developing his landmark silver nitrate technique.

Figure 1-3: Miracle of the Slave. *Tintoretto (1518-1594), one of the last great painters of the Venetian School in the Italian Renaissance, was the son of a fabric dyer, and employed carmine in this 1548 depiction of a Christian slave being saved from torture. It is displayed in the Gallerie dell' Accademia in Florence.*

Another naturally-derived stain used in microscopy in the mid-nineteenth century was hematoxylin, obtained from the heartwood and sap of the logwood, a flowering tree found in Central America, southern Mexico and the Caribbean. (As an aside, the tree was a valuable source of dye for coloring fabrics black, blue or purple in the 17th century, and the struggles around obtaining logwood and shipping it to Spain are deeply enmeshed in the history of piracy.)

c
d
d
b
c
e
a
b
Hæmatoxylum campechianum L.

Figure 1-4: *Drawing of the flowers and pods of the logwood,* Haematoxylum campechianum, *by Adolphus Ypey, 1813. Technically it is a legume, in the pea family* (Fabaceae). *At the time of the discovery of its use by native peoples in what is now Mexico by Cortes and other Spanish explorers, the clothing of common people in Europe was mostly gray and brown; the introduction of dyes made from logwood and other sources made it possible for the first time to inexpensively change the coloration of garments.*

Robert Hooke's *Micrographia* in 1665 (described in the previous section) included material on hair and wool colored with logwood derivatives, and Heinrich Waldeyer, who played an important role in formulating the neuron doctrine (Chapter Three), was the first to use it for microscopy of human tissue in 1865. It stains cell nuclei dark blue or purple, and to this day is widely used in combination with the synthetic dye eosin, discussed in the next section.

Synthetic dyes:

Two years before Joseph von Gerlach entered an apothecaries shop to inquire about dyes that might delineate blood vessels, William Henry Perkin, a 15-year-old chemistry student in London, was engaged in a different sort of project. This was during the time of the Crimean War, in which many soldiers succumbed to infectious diseases including malaria. The only effective treatment at the time was quinine, derived from the bark of the cinchona tree in South America, and there was a growing effort to find a way to synthesize it. During Easter vacation in 1856, Perkin, working in his attic laboratory at home, was trying to do so by oxidizing a complex coal tar derivative, and then aniline, with potassium dichromate. Sadly, because of toluidine impurities in the preparation, he was rewarded only with a black sticky substance which firmly adhered to his glassware. When he attempted to remove it by dissolving it in alcohol, he was surprised to find that it turned into a liquid which stained his cleaning cloths with a deep purple color.

Perkin, who had an interest in photography and painting, was quick to realize that he had stumbled upon a dye with potential commercial value. When his professor discouraged this line of reasoning, urging him to return to his studies, he set up a small laboratory in his back yard, and developed it as a fabric dye, aniline purple (or mauveine) which became remarkably successful after the French Empress Eugenie and Queen Victoria were seen wearing gowns in the new vivid color. In effect, Perkin had founded the synthetic dye industry, and more broadly, the field of synthetic organic chemistry.

As time went on, the manufacturers of synthetic dyes applied their skills to other areas, notably creating medicines, as described in detail in Mendelson (2020). Another pathway led to developing dyes that were effective as histological stains. In 1876, one year before Cajal was to visit Madrid and discover the possibilities of microscopy, Heinrich Caro (1834-1910), a German calico printer and chemist, synthesized methylene blue. It was later used in the 1880s by Paul Ehrlich (1854-1915) to stain the *Plasmodium* parasites causing malaria, and by Robert Koch (1834-1910) to visualize the bacterium responsible for tuberculosis. Ehrlich had also noted that it stained neuronal axons, but did not follow up on the finding. In 1891 he also found it somewhat effective in treating malaria, making it the first fully synthetic drug.

In 1896, Cajal, facing criticism that all his claims were based on a single methodology derived from Golgi's silver nitrate work, set about doing his studies over again with a different stain, and chose methylene blue. He reproduced all his important findings, providing strong support for the neuron doctrine which he espoused (Chapter Three). Particularly important was his visualization of small spines on dendrites, which he had described on Purkinje cells in his first publication using the Golgi method. With time dendritic spines were to take on more importance, as they came to be recognized as the sites of the majority of excitatory connections between neurons (Glickstein, 2006).

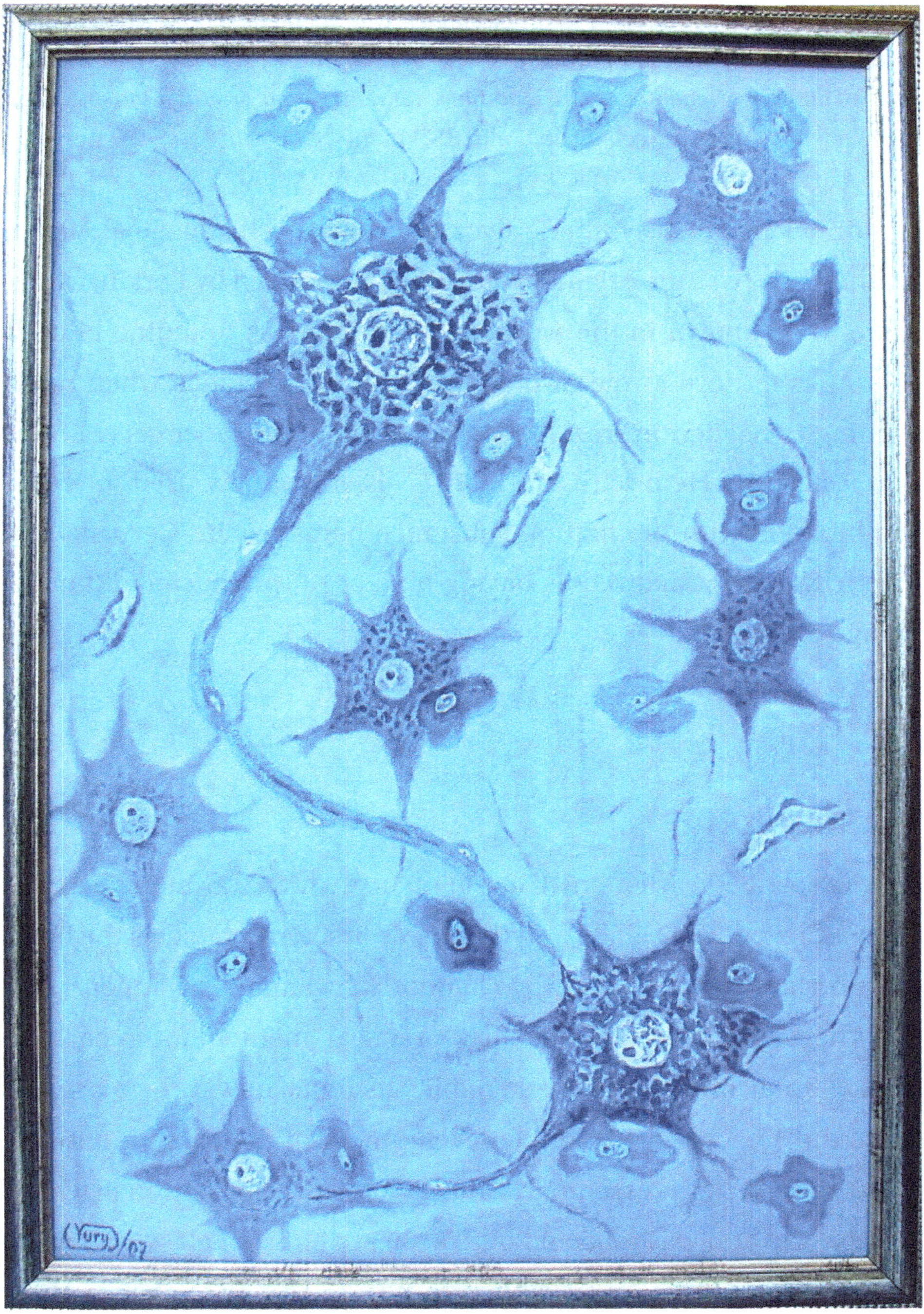
(Yury)/07

Figure 1-5: *Spinal cord motor neurons stained with methylene blue. This process, first developed by German psychiatrist and neuropathologist Franz Nissl in 1894, uses alkaline dyes such as methylene blue, toluidine blue or cresyl violet, which interact with DNA in the cell nucleus and RNA in the cytoplasm.*

A number of other synthetic dyes also became histological stains. Among these was the original mauveine synthesized by Perkin, which came to be known in the world of microscopy as toluidine blue. It continues to have a role in modern times, for instance when rapid staining is needed in frozen sections of tissue from surgery. Eosin, synthesized by Heinrich Caro in 1871, is still widely used, often in combination with the natural substance hematoxylin. Cells stained with 'H&E' are visualized as having blue or purple nuclei and pink-orange cytoplasm.

Photography

As we described earlier, in 1839 cell theory, which was at the heart of the conflict between Golgi and Cajal, was first formulated as the basis of both plant and animal life by Theodor Schwann. As it happens, this was the same year that the first practical method for recording photographic images was made public. Its roots, however, go back as far as the 17th century with the development of the *camera obscura* ('dark room'). It was essentially a box fitted with a lens, through which light passed and was reflected onto a sheet of frosted glass, on which the image appeared for tracing onto paper. This in turn was often transferred onto canvas or wood. The challenge was to find a better way to record and preserve the image.

Figure 1-6: *A dictionary illustration of a* camera obscura, *circa 1850. In the figure, B= lens, M= mirror, and O= path of light without mirror.*

By the 18th century, it became clear that naturally occurring silver salts turned dark when exposed to sunlight, that this was indeed due to the light and not to heat, and that it could be expressed chemically as reduction to metallic silver. Johann Heinrich Schulze (1687-1744), a German physician and philosopher, found in 1727 that sunlight passing through stencils could reproduce writing in the contents of jars containing chalk and silver nitrate. The writing remained until the bottle was shaken, or until continued exposure to light completely darkened the mixture. In the 1790s, the Englishmen Thomas Wedgwood (1771-1805) and Humphrey Davy (1778-1829) were unsuccessful at using the *camera obscura* to preserve images on silver nitrate, as the light was not strong enough, but later developed a technique in which light passing through a painting on glass could be reproduced on a silver nitrate-coated surface, a form of contact printing. This was a major step forward, but still left the problem of fixing, or arresting, the process, so that the plate did not turn completely black.

In the 1810s, Joseph Nicéphore Niépce (1765-1833), a former officer in Napoleon's army and the co-inventor with his brother of an early internal combustion engine, became interested in printmaking. Lithography was becoming popular, but the area in eastern France where Niépce lived as a gentleman-farmer was devoid of the types of stone it required, and he sought alternative techniques. He developed a process which he called heliography ('sun drawing'), in which prints were placed onto metal plates coated with bitumen of Judea, a form of asphalt. The portions of bitumen which were exposed to sunlight hardened, and remained in place when the rest was rinsed off with oil of lavender. The uncoated areas were then treated with acid and aquatint so that they could hold and print with black ink. In 1825 he made the first surviving print, showing a 17th century engraving (Figure 1-7).

Figure 1-7: *Niépce's heliograph of a man leading a horse, now in the Bibliothèque Nationale de France.*

In 1826, Niépce adapted his system to the *camera obscura* using paper coated with silver muriate or nitrate. His original picture required at least eight hours of exposure (though some researchers estimate as long as three days), and represented the scene outside his window. He had, then, created the first fixed image from nature, as well as the first photomechanical system of printing.

While Niépce was making these advances, a Parisian of a very different temperament and style was also interested in recording images. Louis-Jacques-Mandé Daguerre (1787-1851) had a background in architecture, panoramic painting and stage design, and was by nature a showman. In 1822 he and panorama painter Charles Marie Bouton (1781-1853) created the first theatrical diorama, opening in Paris. The audience, sitting in a cylindrical room, would see huge multi-layered sheets of semi-transparent linen, on which paintings had been made on both sides. Manipulation of the angle and intensity of lighting would alter the audience's view of which paintings were most evident, giving the impression of movement. After a 10-minute showing, the entire room containing the audience of perhaps 350 would be rotated on a turntable in order to view a second scene. Typically, one would be an interior, while another might be a landscape. One particularly popular view was a depiction of the Rosslyn Chapel in Scotland, which by legend appeared to burn in a fire which did not consume it whenever a lord of the castle died. (Rosslyn Chapel was later to gain notoriety as a setting in Dan Brown's novel *The Da Vinci Code.*) The diorama was initially a remarkable success, but was declining in popularity when the theater burned down (in this case, by a fire which did consume) in the 1830s, leaving Daguerre without a way of making a living but salved by an apparently handsome insurance settlement.

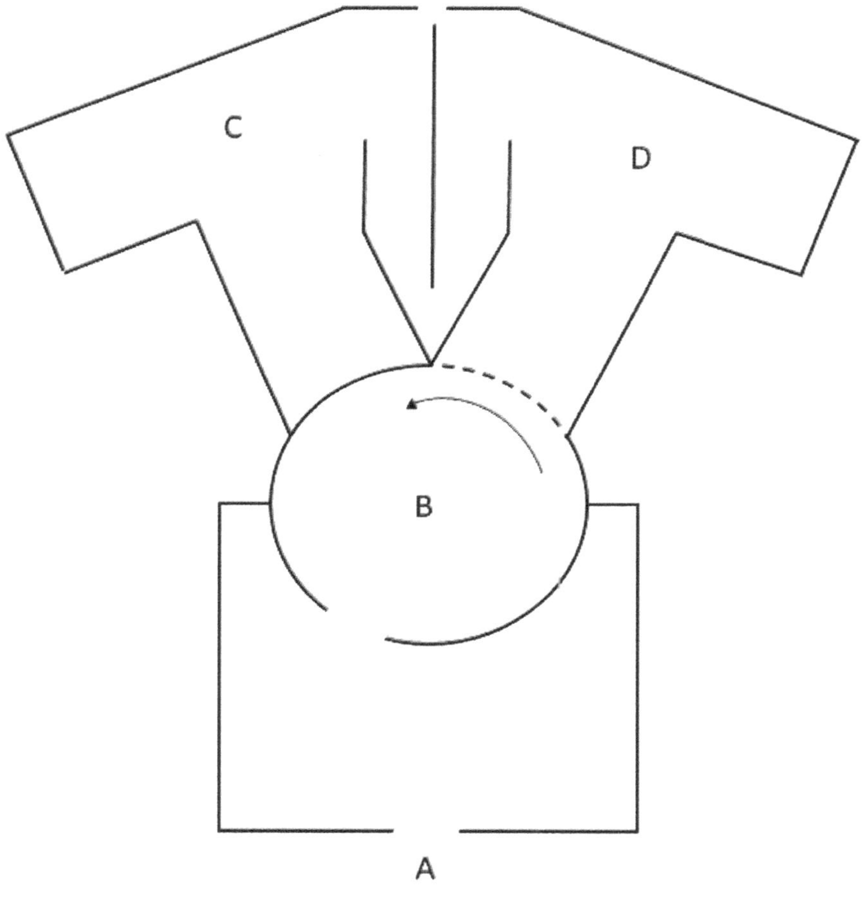

Figure 1-8: *Floor plan of a diorama theater in London, 1823. From the front entrance (A), the audience went into the auditorium (B), which could be rotated so that they could sequentially view two different scenes, set up in chambers C and D.*

Given his profession, Daguerre became interested in capturing images, and explored the possibilities of the *camera obscura* using silver chloride. In 1826, while visiting an optical shop which supplied him with lenses, he learned about Niépce, and was quick to write him suggesting they work together and share the likely profits from the endeavor. Niépce was cautious about such an arrangement, but by 1829, they entered into an agreement. They met in person only once (as did Cajal and Golgi; Chapter Four), but maintained a voluminous correspondence. Both were wary of their techniques being exposed, and devised a numerical code to keep papers secure.

At first the two refined the bitumen process, but were unable to get around the extremely long exposure times which it required, and turned their attention to silver salts. As it happened, Niépce unexpectedly passed away in July 1833. His family's fortune was so diminished that his burial was at public expense. Daguerre continued the work, exposing a silver-coated copper plate to iodine vapor, producing silver iodide. This, too, required a prolonged exposure to produce a visible image, but Daguerre guessed rightly that a much shorter time might produce an invisible 'latent' image, which could be developed into a visible picture. He did so by exposing the plate to mercury vapor, and learned that the process could be halted ('fixed') by washing off the remaining silver iodide with concentrated salt water or sodium thiosulfate. The resulting 'daguerreotypes' appearing in 1837 were to become the first practical, and later commercially successful, photographic process.

Into this story now entered a third figure. Dominique François Jean Arago (1786-1853) was a mathematician and physicist whose first appointment at age 18 had been as secretary to the Observatory of Paris. Two years later he had been assigned to determine measures of longitude in Spain and Majorca, only to be arrested as a suspected spy during this period of French incursion into Spain. He had escaped and been re-captured by a Spanish privateer, imprisoned in a lighthouse and eventually was able to return to Paris. He then devoted himself to science, and later politics. He contributed to the understanding of electromagnetism, wave theory and the polarization of light, as well as astronomy, where his work gave the impetus to search for the planet Neptune.

When Daguerre began showing the results of his work (though withholding the details because of the commercial possibilities) to well-known scientists, Arago, then a permanent secretary of the Academy of Sciences, was very much impressed. By this point, he had moved into politics, and was a member of France's Chambre des Députés. Believing that this knowledge should be available to the world, he negotiated an arrangement. Daguerre, who was having difficulty finding financial backing, would receive a lifetime pension, and Niépce's estate a smaller annual amount, much to the unhappiness of his son Isidore, who believed that Daguerre was unfairly benefitting from his father's work. In exchange, the government would own the rights to the daguerreotype, with the intention of making it public.

In 1839 at a joint meeting of the French Academies of Sciences and the Arts, the work and a general description of the method were revealed. They were widely publicized, leading others including the Englishman Henry Fox Talbot (1800-1877) to describe similar techniques which they had been working on around the same time, but the result was that for the first time the practical creation of images gained international recognition.

Figure 1-9: Daguerre's image of the Boulevard du Temple, in 1838-1839. It is the first *known photograph depicting a living person. Because the exposure time was about 10 minutes, the moving traffic and people walking are not seen. Fortunately, in the lower right (marked by arrow), a gentleman held still for some minutes while his boots were being polished, and his image was preserved.*

As the next few years went by, the introduction of new chemical processes and lenses made shorter exposure times for daguerreotypes possible, though other challenges remained. They needed to be lit and viewed at certain angles, and were so delicate that they required mounting under glass. Moreover, the technique produced only one-of-a-kind images, which could be reproduced only by taking a new picture. In 1841 Henry Fox Talbot introduced the 'calotype' or 'talbotype' system using silver-iodide coated paper, producing translucent negatives, which though of lower quality, could be used to make positive prints easily by contact printing. The 1850's saw the development of the wet collodion process, which after initial use for single images led the way to easier printmaking with higher resolution.

By the 1860's these newer innovations had largely replaced the original daguerreotypes, just as they, too, would later be replaced by the dry plate in the 1870s and later, film negatives, and then digital techniques. Photography had become very popular, with a burgeoning literature available to scientists who might wish to record images. Joseph von Gerlach (discussed in the earlier section on staining and Chapter Two), was one of the first to photograph microscopic images shortly after daguerreotypes first appeared, and wrote a textbook on the subject in the 1840s. In 1872, Charles Darwin published the first book to use photographic illustrations, *The Expression of Emotions in Man and Animals*. In that same year, when Golgi was working on the visualization of neurons in his kitchen laboratory (Chapter Two), the boom in photography had led to the easy availability of silver salt solutions such as the silver nitrate he employed in his groundbreaking discovery reported in 1873.

Summary

By the time Golgi and Cajal began their studies of the histology of the nervous system in the 1870s and 1880s respectively, several technologies had matured to a point facilitating new discoveries. The advent of the achromatic microscope in the 1820s led to advances in the life sciences, including the development of cell theory in the late1830s. In the 1850s with the work of Joseph von Gerlach and others, carmine and other natural stains came into more widespread use, and the birth of synthetic organic chemistry in the fabric dye industry led to the development of synthetic stains such as methylene blue in the 1870s. Photography, born in the 1820s, became practical and very popular by the 1860s, providing access to silver salt solutions, and ultimately a new method of recording microscopic views. Let's turn, then, to the lives of Golgi and Cajal, and how they came to creatively take advantage of these new developments .

REFERENCES for CHAPTER ONE

British Museum, Object type inlay, Museum number 90959. Accessed August 28, 2022A.
https://www.britishmuseum.org/collection/object/W_-90959

British Museum: Collection Items: *Micrographia* by Robert Hooke, 1665. Accessed August 30, 2022B.
https://www.bl.uk/collection-items/micrographia-by-robert-hooke-1665

Glickstein, M.: Golgi and Cajal: the neuron doctrine and the 100th anniversary of the 1906 Nobel Prize. Current Biology 16: R147-R159, 2006.
https://www.cell.com/current-biology/pdf/S0960-9822(06)01203-6.pdf

Mendelson, W.B.: *Molecules, Madness and Malaria: How Victorian Fabric Dyes Evolved into Modern Medicines for Mental Illness and Infectious Disease.* Pythagoras Press, 2020.
https://www.amazon.com/dp/B088QPG14K

Virchow, R.: *Cellular pathology as based upon physiological and pathological histology*. University of California Libraries, 2015 (original 1863).

CHAPTER TWO: CAMILLO GOLGI AND THE RETICULAR THEORY

Life and times

Camillo Golgi came from a family with roots in the city of Pavia, near Milan. His father, Alessandro Golgi, the son of a hat maker, went to medical school there, and married his cousin, the daughter of a carpenter. After graduating in 1838, Alessandro found a position as municipal doctor in the village of Corteno, in a valley among the Lombardy Alps. It was there that Camillo Golgi, the third of four children, was born in 1843. Although the father remained in Corteno, the family moved back to Pavia so that the children could benefit from its schools, and he was able to join them in 1858 when he found a position in a chronic hospital nearby. As Golgi did not write an extensive memoir, there is little known about his childhood, though the records indicate that he was a diligent student, with high marks in all subjects except German. These were the years in which Lombardy was under the control of Austria, leading to a series of rebellions as well as a war between the Piedmontese and Austria, resulting in the later unification as the Kingdom of Italy in 1861. Though Golgi was known as a quiet and cooperative student, he was caught up in the passions of the times, and got into trouble for expressing anti-Austrian feelings in school. He was briefly expelled, but later allowed to complete his last year, and enrolled in the medical school of the University of Pavia in 1860.

Golgi's initial intention was to learn to practice clinical medicine. In 1859, the university, which had been closed during the period of political turmoil, had reopened and benefitted from governmental decrees leading to the formation of faculties including medicine and law. It also benefitted from the influx of doctors and scientists who had gone to Berlin and Vienna for training and returned to share their enthusiasm for deriving information and designing treatments from experimental work, rather than from theory. It was in this environment that Golgi began his studies, and ultimately became enamored of research.

Figure 2-1: *Most of Golgi's career took place at the historic University of Pavia, seen here in 2021. A teaching program there is mentioned in the edicts of a Frankish king in 825 AD, and it is considered to have been a formal university since 1361, thus one*

of the oldest in Europe. Leonardo da Vinci had studied there in 1510 and a strong anatomy program was put in place by the well-known physician Antonio Scarpa in the late eighteenth and early nineteenth century. It was the only university in the Milan area until the founding of the University of Milan in the 1920s.

Golgi graduated in August 1865 with average grades; surprisingly, his lowest marks were in general pathology, the department he would later chair at the university for decades. His dissertation was on the origin of mental illness, sponsored by his professor Cesare Lombroso (1835-1909), director of the psychiatry program, and well-known for his belief that criminality was an inherited condition accompanied by physical characteristics. He began as an intern at San Matteo hospital in Pavia, where he worked on the medical and surgical services. Though he flunked the physical exam for military service, shortages led to his working temporarily as a civilian physician at a military hospital before returning to his training. During an outbreak of cholera the following year, he was required to take care of patients in a nearby village for two months, and was reported to have done a good job. (We will see later that during another outbreak in the 1880s, Cajal, too, was detoured from his studies to work on cholera.) Golgi was promoted to a position as a 'secondary' physician in 1868. One expression of his early interest in research was evident in 1869: in order to help resolve a debate on whether syphilis could be transmitted by human milk, he purposely injected himself with milk from infected mothers (Gazzetta Medica Lombardo, 1902), though nothing is known of the outcome of his self-experiment.

In those years, Italian academe was becoming swept up in the movement known as positivism. Formulated first in social terms by the French philosopher Auguste Comte (1798-1857) in the 1830s through the 1850s, it emphasized that knowledge should be derived by reasoning from objective observations, instead of, for instance, from intuition or metaphysics. Its application to medicine was facilitated by discoveries of the last few decades. In 1828, Fredrich Wöhler (1800-1882) had created the organic compound urea from inorganic substances, suggesting that organic materials were not physically unique (Mendelson, 2020), and the German physicist Hermann von Helmholtz (1821-1894) had applied the physical principle of conservation of energy to living organisms. Darwin's *On the Origin of Species* had become available in Italian in 1864. In 1852 the Dutch physiologist Jacob Moleschott (1822-1893) and others, suggested that even psychological processes could be explained in terms of physical and chemical activities, without the involvement of supernatural intervention. Rudolf Virchow's belief that cellular processes, and not the soul, induced healing, was beginning to take hold, and there was growing popularity of using microscopy to understand illnesses.

Golgi was clearly influenced by such thoughts, and began working more with Cesare Lombroso, studying brain tissue from autopsies of the mentally ill, and analyzing thresholds of discomfort from electrical impulses in psychotic patients. In time he began to distance himself from Lombroso, whom he felt had fallen prey to psychiatric theories not always grounded in empirical observation and studies; phrased differently, Golgi felt that Lombroso thought of intriguing theories, but then interpreted his observations in light of these presuppositions. Ironically, this is exactly what Cajal was later to say about Golgi's insistence on the reticular theory.

Golgi retained Lombroso's fascination with the human brain, but moved into the intellectual circle of Giulio Bizzozero (1846-1901), a brilliant physician and histologist three years younger than himself. Bizzozero is known for later having coined the term 'platelet' in 1881, as well as describing *Helicobacter pylori* bacteria and showing their association with peptic ulcers in 1892. Golgi began spending time at Bizzozero's Laboratory of Experimental Pathology, where he became part of a group of bright young investigators. Though Golgi's modest origins and quiet manner tended to put him in the background, Bizzozero recognized his potential. They became close friends, living in the same building, and subsequently Golgi married Bizzozero's niece.

Under Bizzozero's guidance, Golgi's work flourished. He began to publish, first with a study of pellagra in 1868 and then a monograph *On the etiology of mental alienations* in 1869, emphasizing the importance of moving psychiatry from a theoretical stance to a discipline founded on empirical observations. That year he won a prize for a study, *On the alterations of the lymphatic vessels of the brain,* that notably involved use of potassium dichromate and silver nitrate, which he later employed in sequence in his breakthrough 'black reaction' stain of brain tissue. He went on to studies of the cerebellum, where he followed up on Rudolf Virchow's work on glial cells, again preparing tissue with potassium dichromate. He also worked closely with Niccolò Manfredi, the son of a well-to-do family whose interests ranged from ophthalmology to the therapeutic benefits of music. The two became close friends, sometimes enjoying weekends when talk of science alternated with musical endeavors in which Manfredi played the violin and Golgi the flute. Their professional work was productive as well, including reports on glial cells among the layers of the retina. Golgi had also become an editor of the well-known medical journal *Rivista Clinica*.

Golgi's work was beginning to have an enthusiastic reception abroad, but his prospects for formal advancement at the university turned out to be poor. Though Lombroso and Bizzozero had induced the. university to let him teach a course in microscopy, he was stuck in the same underpaid clinical position he had held since he completed his internship. By early 1872 his position became even more difficult, as Lombroso had left the previous year to become director of an asylum in Pesaro, and Bizzozero was about to move to Turin. Golgi's efforts to find further financial support resulted only in a small stipend for training in the interaction of plant spores and human health. Later that year, at the urging of his father, he took an examination for the position of chief physician and surgeon at the Pie Case hospital for chronic illness near Milan in Abbiategrasso. It was the same one to which his father had moved from Corteno in 1858, when wanting to be closer to his family in Pavia. Though he won the competition, he asked for a delay in commencing his duties there while he looked for other academic positions. Finally, in June he reluctantly reported for work.

The hospital housed approximately 600 patients with a variety of chronic medical and psychiatric conditions. As chief physician, Golgi was required to live there and to be on call at night. He was assisted only by another physician who was only available in the daytime. There were no research facilities of any kind, and though his contract made brief mention of encouraging anatomical and pathological studies and preserving tissue specimens, the hospital director was unwilling to provide funds for any endeavors besides direct patient care. In response, Golgi set up a microscope and a basic laboratory in the kitchen of his apartment, and pursued his studies. (One is reminded of how seven decades later, the Australian psychiatrist John Cade constructed a laboratory in an unused kitchen of a military hospital, in which he re-discovered the clinical use of lithium for bipolar disorder (Mendelson, 2020)).

Whenever he could find free time, Golgi pursued his studies of sensory and motor mechanisms, while also performing his clinical work and struggling with the hospital director's mandates, which included trying to save money by reducing nutrition for the patients. He also spent time caring for patients in a smallpox epidemic, and ever the researcher, published papers on the effects of the illness on bone marrow and spleen. (As mentioned earlier, as a new physician he had treated cholera patients and been involved in issues of transmission of syphilis, providing background for his later movement from neuroanatomy to the study of infectious diseases.) But for now, his interest remained focused on studying the nervous system. He worked alone, intensely, in what some have referred to as monk-like conditions and manner, but was bolstered by an active correspondence with Bizzozero and Niccolò Manfredi. He continued an interest begun in Pavia, working to modify tissue preparation and staining procedures which would better visualize brain tissue. In February 1873 he wrote Manfredi that he had developed a new technique for staining nerve tissue. At first he had thought it was useful for 'interstitial stroma', but soon recognized that it effectively made neurons visible. Essentially it involved fixing (hardening) the tissue with potassium dichromate for 1-45 days, then 'impregnating' it with silver nitrate for varying durations, leading to the precipitation of silver chromate. As the neurons then appeared as black structures which clearly stood out from the yellow background, the process was known as the 'black reaction' (*reazione nera*).

Silver nitrate was already being employed by histologists for staining the intercellular spaces in various types of tissues, and as mentioned earlier, Golgi had previously employed both potassium dichromate and silver nitrate in his studies of the lymphatic spaces. It apparently occurred to him to extend the use of metal precipitation to fixed nervous system tissue, focusing on the visualization of neurons. His new discovery came from doing so, using both potassium dichromate and silver nitrate in sequence, so that they reacted with each other. At the time, most histological work on the nervous system had involved fixation of the tissue, embedding and slicing it, and then staining with carmine or hematoxylin (Chapter One). This had had two disadvantages. Firstly, it only visualized cell nuclei and to various degrees the cell bodies of neurons, with only partial views of their projections. Moreover, by staining all neurons, it, presented a confusing mélange of densely packed cells. Golgi's method only stained about 1-5 percent, making the view easier to understand, and additionally stained not only the cell bodies, but the dendrites and axons which extended from them.

Golgi published his findings in August 1873 in the *Gazzetta Medica Italiana,* a relatively small journal, though abstracts appeared in more widely read publications. There was little interest or response. Part of this may have been due to the lack of illustrations (which did not begin to appear until 1875), so that initially readers did not have the striking visual experience of seeing neurons and their projections startlingly displayed in black, which so impressed Cajal in 1887. In this first publication Golgi also offered few details of how his studies were performed, perhaps because he was still working them out. There were also a number of other new histological techniques being published in those years, which may have diluted its significance in the eyes of researchers who did not appreciate its importance. The black reaction languished, then, for about 14 years, when Cajal recognized its potential (Chapter Two). After that, and with Cajal's modifications, it became a standard technique in most laboratories by the 1890s.

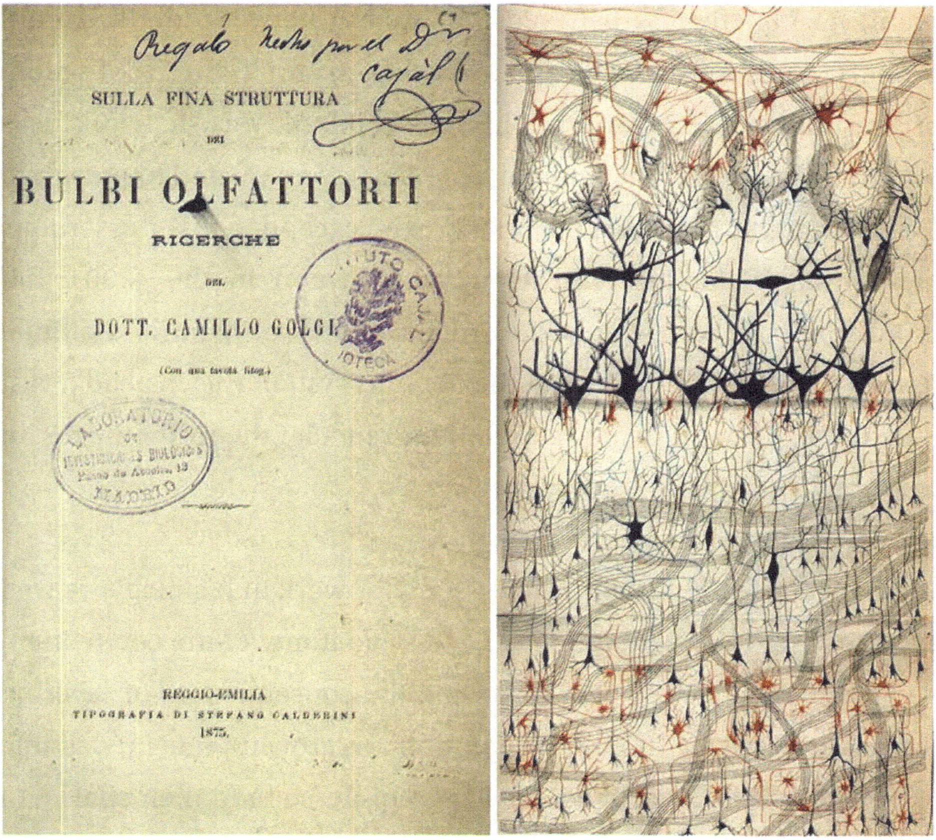

Figure 2-2: *Golgi's first published drawing of tissue using his 'black reaction' staining technique, in 1875, shows a vertical view of the olfactory bulb of a dog, reproduced by DeFelipe (2010), who refers to it as 'semi-schematic'*

In describing the results from his technique, Golgi went on to suggest that axons of cortical gray matter neurons do not necessarily extend into the white matter, but that instead the projections of many quickly branch into smaller fibrils, leading to continuity of nerve cells. Though his studies of axons differed from earlier accounts suggesting that dendrites were fused together in a diffuse network throughout the brain, he nonetheless believed that such a network, derived from axonal connections, existed and was necessary for the 'unity of action' of the nervous system. In his conception, this network was multinucleated, and the functions of nuclei were related to nutrition. He was arguing, then, that the brain was an exception to the assertions of Matthias Schleiden and Theodor Schwann in the 1830s that individual cells were the basic structural units of plant and animal tissue (Chapter One), and that the nervous system was instead a vast reticulum. He continued to support this reticular theory for the rest of his career.

Golgi regularly published reviews of major work in histology, as well as his own studies with gliomas and the anatomy of the cerebellum, and a paper with Bizzozero regarding the consequences of severing the sciatic nerve. By 1875 his recognition was growing, and he was able to return to the University of Pavia as a professor and then chairman of the department of histology. Over the years he rose in rank inside the university, also becoming chairman of general pathology in 1881, and later the rector of the university from 1893 to 1896, then dean of the medical school from 1899 to 1991, and once again rector from 1901 to 1909. At the same time, he held political office in Pavia, and had become a Senator of the Kingdom of Italy in 1900.

Golgi's rise at the University of Pavia reflected his fruitful research work. In 1880 he described the Golgi tendon organ (the branching ends of sensory nerves in tendons) and in 1883 the cytoplasmic organelle which became known as the Golgi apparatus. Around 1885, his interests returned to infectious disease, and by 1887 he had largely left active research in neurohistology. During the next eight years he focused on malaria, and demonstrated that the intermittent nature of its fevers was related to the timing of *Plasmodium* parasite spores exiting red blood cells and entering the circulation. He was engaged in this work when Cajal became aware of the black reaction and devoted himself to improving it and understanding its implications for neuronal function.

Figure 2-3: *Camillo Golgi, in approximately 1903.*

The Reticular theory

In 1815, the French physiologist Marie Jean Pierre Flourens (1794-1867), who is often considered to be one of the fathers of experimental brain research, began to deal with the question of whether various functions of the nervous system were localized to specific areas. This notion had been raised by the Austrian Franz Joseph Gall (1758-1828), and expressed in a rather haphazard and speculative manner in his development of phrenology. Flourens approached this question by making lesions of the nervous systems of rabbits and pigeons, and reporting on their changes in behavior. Lesions of the cerebellum in rabbits, for instance, led to loss of equilibrium, while damaging the horizontal semicircular canals in the vestibular system of pigeons caused them to fly in circles.

Over time, then, Flourens recognized that certain parts of the brain were indeed associated with various activities. He believed that the cerebral hemispheres regulate perceptions, some measures of judgment, and motor behavior. In contrast, the cerebellum seemed related to motor coordination, while the medulla of the brainstem was associated with vital functions such as breathing and blood circulation. He was, however, unable to pinpoint areas for memory and cognition, in retrospect apparently partly due to the difficulty in studying them and partly due to the relatively less developed cerebral cortex in pigeons. For this reason, he believed that they were derived from a diffuse system widely distributed in the brain. This 'globalistic' conception was to grow, seemingly gaining support from microscopic studies, from the 1830s through the 1870s. In the next few paragraphs we will look at some of these studies, and how they may have influenced Golgi's thinking.

One of the first reports of visualizing neurons was in 1833, by Christian Ehrenberg (1795-1876), a German anatomist and zoologist, who described 'ganglionic corpuscles' in leeches. These likely were neuronal cell bodies and portions of their projections. His 1836 book colorfully described his studies as *Observations of a So Far Unknown Structure of the Organ of the Soul of Humans and Animals*. He came to believe that the interconnections were similar to those of the vascular system. His work was quickly followed in 1838 by Jan Evangelista Purkinje (1787-1869), who published the first illustrations of mammalian neurons.

As we described in Chapter One, in 1858 Joseph von Gerlach began using carmine stain to study tissues. He was becoming increasingly interested in the nervous system, and having seen how carmine was absorbed into other types of cells, applied it in combination with gelatin. He visualized cell bodies and noticed projections (later to be recognized as dendrites and axons) but could not clearly see either their attachment to cell bodies or their endings. Unable to effectively trace them, he speculated that they were part of a network connecting all cells in a vast reticulum, a view which he formalized in 1871. In passing, he was also very much interested in photography of microscopic images, using the technique in the 1840a soon after the development of the daguerreotype (Chapter One), and wrote one of the earliest books on medical photography.

In the early 1860s, the German anatomist Otto Deiters (1834-1863) used carmine stain on tissue hardened with potassium dichromate, and then painstakingly separated cells with a needle while viewing through a microscope. He was able to better visualize neurons as well as axons ('axis cylinders') and dendrites ('protoplasmic processes'), though he too was unclear about their attachment to the cell bodies. His method was very labor intensive, however, and was more suitable for studying the spinal cord (Glickstein, 2012). He came to believe that dendrites were physically connected together to form a vast network. Before his drawings could be published, he succumbed to typhus at age 29 in 1863. When another anatomist, Max Schultze, organized Deiters' data and published his figures from spinal cord tissue two years later, his work was taken to support reticular theory.

Golgi's conception of a vast reticulum of anatomically connected cells derived from a deep-seated belief that the nervous system was a unitary entity, a 'nerve organ' linked by axons and expressed most clearly in the cerebral gray matter, which carried out the fundamental processing in the brain. He continued to assert that signals did not travel along specific pathways, but rather were broadcast as electrical impulses to the entire nervous system. He theorized that such propagations were not entirely random, but that there could be some selective fields of neurons which might be more sensitive to particular propagations. In general, though, Golgi believed that specific transmissions along pathways to specific areas could not explain the complex function of the brain, which could be better understood if the information were available to the entire system. It seemed unlikely to him that a structure built on a multitude of individual elements would be capable of integrating information. He did not appear to understand the consideration raised by the French philosopher Denis Diderot (1713-1784) a century earlier, that a situation in which all incoming information was infiltered and flooded the brain could lead to disarray, a kind of pandemonium expressed as madness.

Golgi also did not appreciate that the reticular view could not easily account for the specific activities of different regions of the brain, coming to light, for instance, by the studies of the cerebral cortex by the French physician Paul Broca in 1861, as well as by the German anatomist Gustav Theodor Frisch and neurologist Eduard Hitzig in 1870. Similarly, it was not well equipped to explain the temporal progression of many nervous system processes including successive thoughts as formulated by the Austrian psychiatrist and neuroanatomist Theodor Meynert (1833-1892), who incidentally was also Sigmund Freud's teacher (Mendelson, 2022). These seemed to require a system for making signals unidirectional, with gating mechanisms regulating the transmission, and specific pathways. Though Golgi may have speculated that the differential sensitivity of various regions was due to differing densities of the network, these processes were much more easily explained by the neuron doctrine.

REFERENCES for CHAPTER TWO

DeFelipe, J. (2010). *Cajal's Butterflies of the Soul: Science and Art.* New York: Oxford University Press.

Gazzetta Medica Lombarda, p. 431-440, 1902.

Glickstein, M.: Golgi: a life in science. Brain 135: 301-303, 2012. https://academic.oup.com/brain/article/135/1/301/323788

Mendelson, W.B.: *The Curious History of Medicines in Psychiatry.* Pythagoras Press, 2020.
https://www.amazon.com/dp/B083ZRMCW1

Mendelson, W.B.: *Molecules, Madness and Malaria: How Victorian Fabric Dyes Evolved into Modern Medicines for Mental Illness and Infectious Disease.* Pythagoras Press, 2020.
https://www.amazon.com/dp/B088QPG14K

Mendelson, W.B.: *The Psychoanalyst and the Nazi Nobelist.: the Curious Story of Sigmund Freud and Julius Wagner-Jauregg.* Pythagoras Press, 2022. https://www.amazon.com/dp/B09X6BKN4V

CHAPTER THREE: SANTIAGO RAMÓN Y CAJAL AND THE NEURON DOCTRINE

Life and times

Santiago Ramón y Cajal was born in 1852 in Petilla, a tiny village accessible only by foot and muleback in a remote area of Aragón, in northeastern Spain. Its only claim to fame at the time was as the site of the 15th century battle depicted in Verdi's opera *Il Trovatore,* in which Count di Luna and the troubadour Manrico, unaware that they are in fact brothers, fought in opposing armies. This unpromising background was to haunt Cajal for many years; he found it difficult to be accepted in the intellectual circles of Spain's larger cities, just as later coming from Spain made it more challenging for him and his work to be accepted in the more prestigious capitals of European science at the time.

Figure 3-1: *Petilla de Aragón (marked by arrow), with the Pyrenees Mountains in the background. In the nineteenth century, it was so isolated that, according to his biographer, even as late as when Cajal visited it as an adult in 1892, the villagers had never observed a wheeled vehicle (Ehrlich, 2022). In 2008, a classification of cortical interneurons, the 'Petilla Terminology', was named after his birthplace (Ascoli et al., 2008).*

Santiago's father, Justo Ramón y Casasús, came from a family of poor farmers, and had seemed destined to follow in their footsteps, but in an effort to improve his prospects of marrying Antonia Cajal, a weaver's daughter, he became the apprentice of a barber-surgeon. Over the years he continued his training and provided healthcare while the family moved through a series of villages, often after quarreling with the administrators, until settling in Zaragoza (Saragossa), the capital of Aragón. He finally earned a medical degree, and became known as Don Justo, when Santiago was six, and found a post as anatomist at the University of Zaragoza.

Don Justo's lack of an early education and the long struggle to complete his training led him to emphasize schooling for Santiago. It was an effort which was made especially difficult by the child's headstrong and anti-authoritarian nature, leading to beatings, as well as a frequent change of institutions. Perhaps the climax of his misbehavior was when he built a cannon, with which he blew up a neighbor's gate, resulting in several days in jail. Around 1865, about age 13, he devised a scheme to cross rooftops to filch sweets from the house of a neighbor who was a confectioner. In the process, he found that it also contained an extensive library, and he began to 'borrow' books which opened up a world of adventure, including *The Three Musketeers, The Count of Monte Cristo,* and *Don Quixote*. The romance of escapades and travel to new places seemed an antidote to his father's efforts to drill a utilitarian outlook into him.

When he was 14, Santiago began his own literary efforts, writing a guide devoted to the fine art of building and using a slingshot to maximum effect in vanquishing the enemy. He was also to go on to write a novel based on the story of Robinson Crusoe. He produced the illustrations for both these tomes himself, and it became clear throughout his youth that he was attracted to drawing. His father, wishing him to become a doctor, discouraged his interest—even going so far as to pay an artist to say that he had no talent. His father tried a number of stratagems, including apprenticing Santiago to a barber and then a cobbler to learn discipline. At another point he had Santiago join him while he cared for injured soldiers during one of Spain's many nineteenth century insurrections. In 1868, he began to take him for clandestine nighttime visits to the public ossuary at a nearby graveyard, where the bones of former occupants were collected when their resting places were removed to make way for new graves. They carried the bones back home, where Don Justo taught him their anatomy and the necessity for close observation, while the son's drawings of them impressed his father and led to at least a partial reconciliation. Indeed, Santiago enrolled in medical school at the University of Zaragoza that year, graduating at age 21 in 1873.

Shortly after graduation he was drafted, and sent with the army assigned to quell the rebellion in Cuba. There he contracted malaria, had a medical discharge, and came home ill and frail. After some months of recovery, he began seeing patients in his father's practice, but he was considered aloof. It is possible that to some degree this reflected his own personal manner, but it also may have reflected that his first significant clinical experience had been in the wartime setting in Cuba. In any event, he made it clear that clinical medicine was not for him, and in 1876 his father helped him find positions teaching anatomy in a provincial medical school.

In 1877 Don Justo finally achieved his dream, getting his medical doctorate as a full-fledged physician. As having a doctorate was the most basic step in achieving success in mainstream academe, he pressured his son to earn it as well. After taking the preparatory courses in Zaragoza and writing a dissertation on the topic of inflammation, Cajal traveled to Madrid for his examinations. He passed them with unremarkable grades, and while there visited Aureliano Maestre de San Juan (1828-1890), the leading Spanish histologist and a judge in the examinations. It was an experience which was to change the course of his life.

Cajal had known about microscopy in principle but had previously had little hands-on experience. Maestre's enthusiastic assistant Leopoldo Lopez Garcia taught Cajal the use of the instrument as well as the application of dyes such as the red pigment carmine (Chapter One) to stain cells. The colorful views of microscopic tissues caught Cajal's imagination, and perhaps tapped into his fascination for exploration and adventure which he had experienced since childhood, and which may have fueled his unfortunate foray to Cuba. He was quick to buy a French Verick microscope using his discharge money from the military, and set up his own laboratory in his father's attic. There he eagerly began to explore all the body's cells, and wrote a number of papers under the title of 'The Marvels of Histology'. In them he conveyed his enthusiasm, but also, to his later regret, advocated spontaneous generation of living tissue. In doing so he ran afoul of the belief that 'all cells come from cells', the central thesis of the cellular theory of life (Chapter One) which he and mainstream biology came to accept.

Cajal's work in those years was interrupted by two events. The first was the necessity of taking the 'oposiciones', oral and written examinations for lifetime government positions and hence financial security. Much of it depended on memorization of extensive verbal material and rhetoric, neither of which were his strengths—his remarkable memory was more visually oriented, and he was at best an average orator—and he was unsuccessful. His health had never been good since his return from Cuba, and not long after this disastrous result he developed fever and began coughing up blood. He was diagnosed with tuberculosis and spent several months in bed at home and then later stayed at a spa in the Pyrenees. His mood deteriorated, and he had thoughts of suicide as he spent time writing poetry with dark themes. As time went on, his physical symptoms improved. He became more active, apparently believing that this was made possible by the force of his own will. By October 1878 he had returned to Zaragoza, feeling much better, and after a few months serving as a village physician at his father's insistence, went back to his attic laboratory.

With the help of his father, in March 1879 Cajal got a low-ranking faculty position and appointment as director of the Anatomical Museum in Zaragoza, with a salary that made it possible to live separately from his parents. That spring he met a young lady whom he had briefly seen in his schooldays, and by summer they had married and moved into a modest apartment which also served as his laboratory. There his work flourished, marred only a few years later by his belief that he had seen a thread-like structure that seemed to stitch together epithelial cells, a picture of which adorned the cover of his first textbook. Later, using better equipment, he realized that this 'finding', which would be more consistent with the reticular theory of continuous connection between cells, was in error, and removed the erroneous image.

In 1883 Cajal took the 'oposicion' exam for the third time, and passed with high marks, landing him a position in Valencia in January 1884. There, as a full professor for the first time, he prospered in the environment of a first-rate university in a culturally rich town. He was free to pursue his research, though with a brief detour in 1885 into bacteriology when officials in Zaragoza asked for his help during a cholera epidemic. He confirmed that the causative agent was the comma-shaped *Vibrio* bacterium—still not fully accepted by many doctors, who thought 'bad air' was responsible and remained doubtful of a bacterial basis for the illness—and evaluated a controversial vaccine that had been developed by a Spanish physician. Returning to his original work, in the next few years he published a series of updates of his *Manual of Histology and Micrographic Technique*, covering most of the tissues of the body, with one glaring exception—the nervous system. Then came the breakthrough.

Figure 3-2: *Ramón y Cajal, self-portrait in Valencia, 1884-1887.*

In January 1887, Cajal went to Madrid as a judge in the oposiciones, and while there met Luis Simarro Lacabra (1851-1921), a neuropsychiatrist who had recently trained in Paris under the famed neurologist Jean-Martin Charcot (1825-1893) and histologist Louis-Antoine Ranvier (1835-1922). Among the things Simarro showed him in his laboratory were slides he had brought from Paris, of brain tissue stained using the 'black reaction' method developed by Camillo Golgi in 1873. Cajal was fascinated by the sight of neurons well-defined in black, standing out from a yellow background. He likened their sharpness and clarity to the black lines in Chinese ink drawings, and unlike preparations with carmine and hematoxylin, they more clearly showed both cell bodies and their attached processes.

Figure 3-3: Portrait of Dr. Simarro at the Microscope, *1897, by Joaquín Sorolla (1863–1923). He had returned from Paris in 1885 and was in private practice in Madrid when he showed Cajal slides using the Golgi technique in 1887, changing the course of Cajal's career.*

Simarro himself, like most histologists of the time, was very skeptical of the Golgi method which was now 14 years old. It was thought to be unreliable and somewhat capricious, requiring modifications with different tissues and even with the time of the year. Many had grown frustrated with its seeming inconsistency: when tissue samples were stained by the Golgi technique, one sample might work while the next did not, necessitating the preparation of dozens of slides in order to get one just right. Cajal, though, recognized the method's potential. He set out to refine it, and then use the improved technique to clarify the structure of neurons and their anatomical relationships to each other. In 1888 he became the chairman of the histology department at the University of Barcelona, where he thought he might find more international recognition in its cosmopolitan environment, and he continued to expand on Golgi's methods. For example, while Golgi had soaked tissue in silver nitrate for one or two days, he found that two short soakings produced more reliable results.

The density of neurons in brain tissue continued to represent one of the biggest challenges to clarifying their structure; thus, Cajal turned to embryonic tissue, which in early stages was less complex, and which had not yet formed myelin sheaths, to which the silver stain attached poorly. He began with cerebellar tissue, presumably because Golgi had used it, but it was a remarkably lucky choice, as it was particularly suited to histologic study. There, in 1888, using tissue from a three-day old pigeon embryo, he found the large pear-shaped Purkinje cells, surrounded by basket-like processes from stellate cells—in close proximity, but not touching. He also noted a multiplicity of small spines on the dendrites of the Purkinje cells, which later turned out to be an important point of communication. Cajal went on to look at many other parts of the nervous system, coming to similar conclusions.

Cajal created at his own expense *The Trimonthly Review ('Revista') of Normal and Pathological Histology* to publish his results, and in May 1888 printed what he believed was clear evidence that neurons were independent cells. He only had funds to print 60 copies, which he sent with covering letters to various scientists. After this and two more subsequent issues, he received no responses. It seems likely that many of the recipients did not read Spanish, and may not have known what to make of a histologist working in Spain, which until then was not as well established in the international scientific community. In 1889 he sent a copy of his crucial paper in *The Trimonthly Review* directly to Golgi, who responded with a paper on malaria, on which he was currently working, not having been actively involved in neurohistology for some years (Ehrlich, 2022, p. 179).

In 1889, Cajal also attended an international meeting in Berlin, at which Golgi was not present, but Cajal was able to attract the favorable attention of Albert Kölliker (1817-1905), a Swiss histologist who was one of the leaders of the field. In earlier correspondence Cajal had suggested that he might visit Golgi after the meeting. He did travel to Pavia, but was not able to see him. Whether Golgi simply declined, or whether he was gone on vacation was not clear, but the message seemed to be that Golgi did not think a confrontation with Cajal was worthwhile.

Undeterred, Cajal continued his work. He published a description of collateral fibers branching from axons in embryonic chicken tissue, which upset Golgi, who had made a similar finding years earlier. It seems unlikely that either Cajal or Kölliker knew about Golgi's finding at the time, as it had only been published in an obscure Italian journal. Nonetheless Golgi was puzzled—why all the interest in the work of a much younger man who for only two years had been studying the nervous system using his methodology?

Cajal also began to look at embryologic tissue across time. As we mentioned earlier, his original interest had been in its relative simplicity and lack of myelinization, but he also recognized that it provided an opportunity to see how neurons grew and developed. In 1890, as he watched axons protrude, he came to believe that instead of following preordained paths, projections from retinal cells moved along inside a 'growth cone', sometimes hesitating, sometimes changing direction until they reached proximity to other cells. He later came to speculate that axons are guided in their growth by increasing concentrations of chemical signals from other neurons (chemotaxis).

In 1892, Cajal moved to Madrid, assuming the chairmanship of histology that had been held by his former mentor Aureliano Maestre de San Juan. Now recognized as the leading histologist in Spain, he focused his attention on additional ways in which he might convince others of his findings. Much of the international scientific community remained skeptical of Cajal's work, and one criticism was that it was based entirely on the Golgi method, albeit greatly improved. To counter this argument, Cajal reasoned that there would be greater acceptance if he could replicate his findings using an entirely different stain. He chose methylene blue, which as described in Chapter One was created in 1876 by the German calico printer and chemist Heinrich Caro, and was a product of the rapid growth of synthetic organic chemistry with roots in the fabric dye industry (also see Mendelson, 2020).

Part of Cajal's choice of methylene blue likely came from Paul Ehrlich's observation that it could stain neuronal axons, and in 1896, he set about using it to repeat his previous work in the cerebellum and other brain regions. He was able to replicate his earlier findings, including the appearance of dendritic spines, which Golgi and others had thought to be artifacts. Insofar as these small protuberances on dendrites were later to be found to be the site of the majority of excitatory connections between cells, this was an important contribution to the neuron doctrine. In order to capture the much more colorful images he visualized, he turned from ink to produce striking pictures from watercolors. And most importantly, he produced the strongest evidence yet that nerve tissue was made up of individual nerve cells which were in close proximity but were not fused to each other.

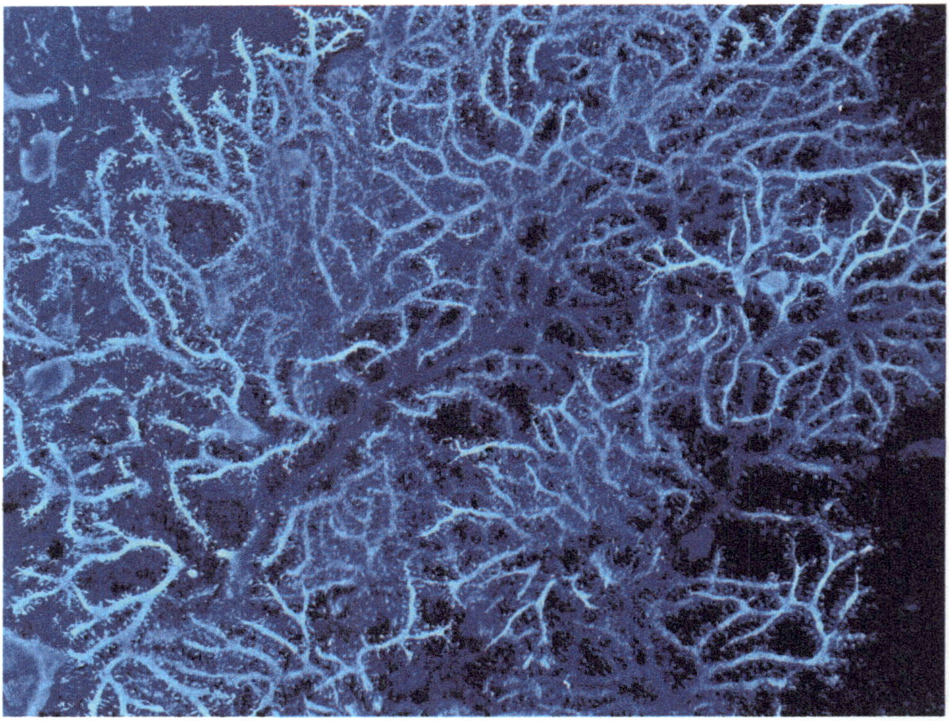

Figure 3-4: *The 'dendritic tree' of a Purkinje cell, on which dendritic spines are visible, in this 2014 scanning electron micrograph of tissue from the rat cerebellar cortex. Dendrites do not merely passively receive signals, but modulate them, making it possible for more complex information to be transmitted. There are a number of thoughts about to the function of spines, including that they increase the area available on dendrites, provide a mechanism for varying the resistance of a synaptic input to the dendrite, or participate in long-term potentiation or depression, processes related to learning (Glickstein, 2006).*

Cajal likened the gray matter of the cerebral cortex to a flower garden and its pyramidal cells to 'butterflies of the soul', which he thought might be the key to mental processes (DeFelipe, 2013). In 1899 he published a major textbook *Textura del Sistema Nervosa,* summarizing his histological studies of the nervous system. He defended the neuron doctrine against a resurgence of a new form of reticular theory advocated around 1901 by the German physiologist Albrecht von Bethe as well as the Hungarian physician and histologist Stephan (Istvan) von Apáthy, whose nerve regeneration studies had led them to suggest that nerve cells were all connected by a network of small neurofibrils. In response, Cajal modified the silver nitrate method to better visualize them, and successfully argued that they were actually a kind of internal skeleton of neurons.

Cajal also began to turn his attention to broader issues, such as how his observations in the microscope translated into the experience of consciousness. He asserted that learning took place by the formation of new connections, and more broadly that experience and changing beliefs might have counterparts in neuronal functioning. For this reason, he objected to the analogy which was often drawn between the nervous system and the newly developed telegraph, which was comprised of a rigid network. He came to believe that the ability of the brain to change with experience was more easily explained by the neuron doctrine than by the workings of a vast reticulum. There was a certain irony insofar as the conception of the neural substrates of how a person could alter their beliefs was one of the major differences with Golgi, who in the face of overwhelming evidence supporting the neuron doctrine, never did change his mind.

Figure 3-5: *Santiago Ramón y Cajal, approximately 1899.*

The neuron doctrine

In 1886-1887, the same time period in which Golgi had moved on from neuroanatomy, a number of reports began to appear questioning reticular theory. Edward Schafer (1850-1935), an English physiologist, who among other things also coined the terms 'endocrine' and 'insulin', studied nerve plexuses in jellyfish, and reported that neurons were separate and not continuous. He thought, however, that they might communicate by an electrical 'inductive action', functionally forming a network. An embryologist, Wilhelm His (1831-1904) in Leipzig, and a psychiatrist, Auguste Henri Forel (1848-1931) in Zurich, produced data suggesting that neurons were the basic anatomic units in the nervous system. His, who also coined the term 'dendrites', examined tissue from human embryos and found no continuity between neurons, though he did not rule out the possibility that they might fuse at a later time. Forel, known for work in other areas including studies of the social behavior of ants, had looked at the effects of sectioning peripheral motor nerves in animals. He found that doing so resulted in death of only a small number of cells in the brainstem, whereas the reticular theory would have predicted much more extensive injury if all cells were fused together. In an 1887 paper he concluded that processes extending from the cell bodies of neurons intertwine, but are not continuous with each other. Cajal always asserted that he had not known about the reports from His and Forel when he published his groundbreaking paper in 1888, which seems plausible in that even these two researchers were not aware of the other's work at the time.

In 1888, Fridtjof Nansen (1861-1930), a Norwegian graduate student, applied the Golgi method to the nervous system of invertebrate marine animals, and in his dissertation published in English he reported no evidence that nerve cells fused together. A few days after completion of his degree, however, he left for an expedition to Greenland, and was the leader of the first group to cross it by ski. He did no further work in neuroanatomy, but went on to a career in diplomacy. Later he was to go on to become the High Commissioner for Refugees at the League of Nations in 1921, and in 1922 won the Nobel Peace Prize for his work with refugees from World War I.

The German anatomist Heinrich Waldeyer (1836-1921), though he did not contribute experimental data of his own to the topic, became very involved in the issue of whether nerve cells were best viewed as individual units or as part of a reticulum. In the process, he went so far as to learn Spanish so that he could read Cajal's papers. Ultimately, he came to support Cajal's views, promoting them in the German scientific literature. In the process, he coined the word 'neuron', derived from the Greek for a sinew or cord. (He had also earlier created the word 'chromosome' in his studies of cytology.) By 1891 he had synthesized the most complete account of the neuron doctrine at the time, and published it in a prestigious German medical journal. The doctrine ultimately matured to consist of three main points: that individual neuron cells are the fundamental anatomic, physiologic and embryologic units of the nervous system.

The neuron doctrine, then, represented a growing, but not fully accepted, point of view. It remained for Cajal to continue to provide more supporting evidence, as well as a conceptual framework for how neurons might communicate. Beginning in 1891 he proposed what came to be known as the principle of dynamic polarization. It suggested that nerve impulses travel from the dendrites to the cell body, and then through the axon to a terminal which communicates across a gap between cells (to be named the *synapse* in 1897 by Charles Scott Sherrington). In this conception, the dendrites and cell bodies of neurons are involved in processing information, while the axon and its terminal play a role in transmission. On a more theoretical level, this may have appealed to Cajal's sense of order, a reaction to what he viewed as a nervous system which would be thrown into chaos if it were as envisioned in the reticular model. As described earlier he had also replicated his studies using methylene blue, and provided a model by which the neuron doctrine could encompass learning and changing beliefs by modification of connections. Ultimately the neuron doctrine was affirmed in the 1950s by the advent of electron microscopic images of individual neurons connected by synapses (Guillery, 2000).

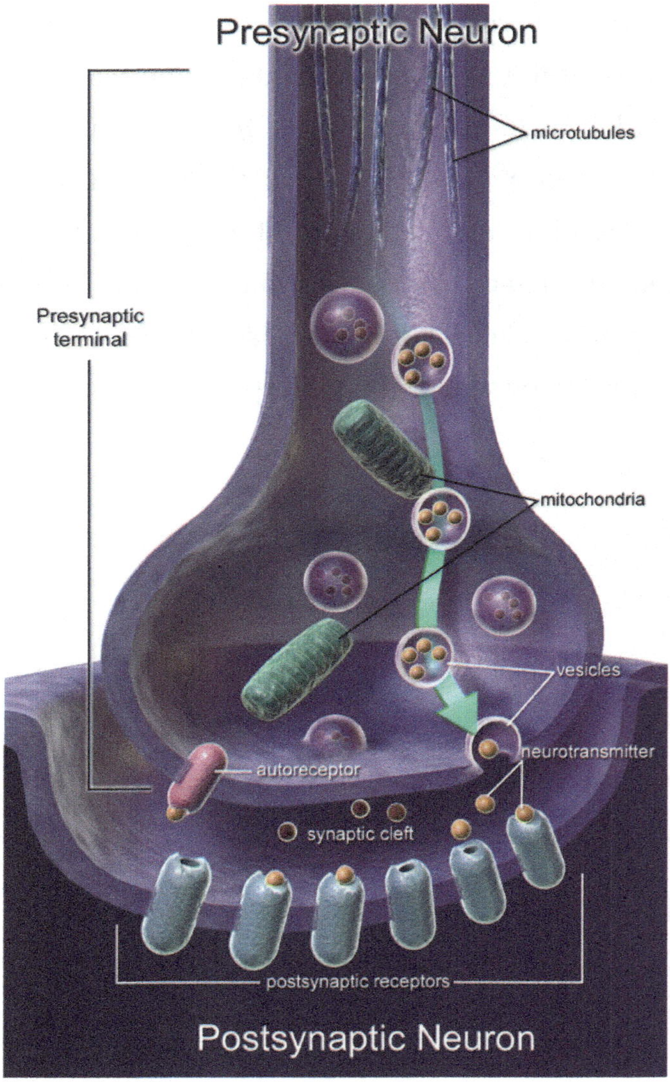

Figure 3-6: *Simplified diagram of a chemical synapse. Vesicles containing molecules of a neurotransmitter in the presynaptic neuron (top), release them into the synaptic cleft, which they cross, binding to receptors on the postsynaptic neuron (bottom) and hence providing a signal between cells. The amount of available neurotransmitter in the cleft can be altered by chemical breakdown, diffusion, or re-uptake into the presynaptic neuron. The mammalian nervous system also contains electrical synapses, which are described in Chapter Five.*

REFERENCES for CHAPTER THREE

Ascoli, G.A. et al.: Petilla Terminology: Nomenclature of features of GABAergic interneurons of the cerebral cortex. Nature Reviews Neuroscience 9: 557-568, 2008.

DeFelipe, J.: Cajal's butterflies of the soul. OUPblog, Oxford University Press's Academic Insights for the Thinking World, Nov. 9, 2013. Accessed July 27, 2022.
https://blog.oup.com/2013/11/cajal-butterflies-of-the-soul-cerebral-cortex/

Ehrlich, B.: *The Brain in Search of Itself: Santiago Ramón y Cajal and the Story of the Neuron.* Farrar, Straus and Giroux, New York, 2022.

Gazzetta Medica Lombarda, p. 431-440, 1902.

Glickstein, M.: Golgi: a life in science. Brain 135: 301-303, 2012.
https://academic.oup.com/brain/article/135/1/301/323788

Glickstein, M.: Golgi and Cajal: the neuron doctrine and the 100th anniversary of the 1906 Nobel Prize. Current Biology 16: R147-R159, 2006.

https://www.cell.com/current-biology/pdf/S0960-9822(06)01203-6.pdf

Guillery, R.W.: Early electron microscopic observations of synaptic structures in the cerebral cortex: a view of the contributions made by George Gray (1924-1999).
https://www.biusante.parisdescartes.fr/chn/docpdf/guillery.pdf

Mendelson, W.B.: *The Curious History of Medicines in Psychiatry*. Pythagoras Press, 2020.
https://www.amazon.com/dp/B083ZRMCW1

Mendelson, W.B.: *Molecules, Madness and Malaria: How Victorian Fabric Dyes Evolved into Modern Medicines for Mental Illness and Infectious Disease*. Pythagoras Press, 2020.
https://www.amazon.com/dp/B088QPG14K

CHAPTER FOUR: STOCKHOLM, DECEMBER 1906

The Introduction of this book describes the climactic awarding of the Nobel Prize to Golgi and Cajal. In summary, it was the first time they had ever met face to face. When Golgi first arrived, he did not take an opportunity to talk with Cajal, who had come to greet him at the train station, and he focused the next day on preparing his talk. Instead of reviewing his own work, Golgi's lecture was spent attacking the neuron doctrine. In contrast, Cajal's lecture was a step-by-step account of his understanding of 'The structure and connexions of neurons', reserving comments for the very end saying that it was sad that Golgi, for all his earlier contributions, was in error.

It may be worthwhile to look a little more closely at Golgi's situation at that point, and the content of his talk. He was 63, at a very different point in his life from when he had discovered the black reaction. He was engaged in the responsibilities of being the rector of the University of Pavia, as well as serving as a senator. He had done little research for the last few years, and even before that he had been focused on a very different area—malaria and other infectious diseases. From the very beginning of his written contact with Cajal, he had had difficulty deciding what to make of him. In the late 1880s he had been puzzled when a researcher with little reputation, who had worked with his method for only two years, wrote challenging letters to him from Spain. His puzzlement later had changed to anger, as Cajal published findings such as the branching of collateral axons, which Golgi had described earlier.

When arriving in Stockholm, Golgi was feeling ill, his wife was in pain from a fall on the ferry, and he was anxious about needing to prepare for his talk. This may have explained his apparent brusqueness with the reception committee, but it also became a lost chance to have some sort of reconciliation, or at least acceptance, with Cajal. Their inevitable dealings during the social events were described as polite. Cajal apparently thought he showed 'Olympic pride and pretentious mien', though it is not clear to what, if any, degree, this was an expression of feelings built up from such a long conflict. Some considered Golgi as uncomfortable and nervous.

Golgi had good reason to be distressed. He was in a confrontation with a much younger adversary who was aggressive and actively involved in research which Golgi had not participated in for years. It has been suggested that he may have felt buoyed by papers coming out since 1901 by Albrecht von Bethe and Stephan von Apáthy (Chapter Three), who had found what they believed to be a network of neurofibrils connecting nerve cells together, which could be interpreted as a kind of revised reticular theory. This seems less likely, as portions of Golgi's lecture were a criticism of this work, which he termed 'hardly conclusive', and at another point he said that this putative series of connections 'has nothing in common with the real diffuse nerve network' (Golgi, 1906). Golgi was sticking by his original conception.

Early in his talk, Golgi proposed that the neuron doctrine was based on three propositions:

1. '*The neuron is an embryological unit*, that is to say, it derives from a single embryonic cell.
2. *The neuron, even in its adult form, is one cell.* The whole system - ganglion cell, protoplasmic and nerve processes - in the adult animal too, is only one cell.
3. *The neuron is a physiological unit...*' (Golgi, 1906).

Golgi took aim at each of these, and explained why he believed they were wrong, though he brought no new data to bear. He was primarily re-stating positions he had formulated more than two decades previously. He argued against Cajal's principle of dynamic polarization—as did others such as the English physiologist Charles Scott Sherrington (1857-1952)—but he relied on his incorrect belief that the role of dendrites was largely nutritional. Overall, his presentation was largely as if the discoveries of the last two decades had not taken place or were irrelevant, as he formulated his notion of 'this network, which I did not hesitate to call a nerve organ'.

A final thought about the Nobel lectures is that both parties made some unusual remarks about the other. As mentioned in the Introduction, Golgi surprisingly commented that the neuron doctrine 'is generally recognized to be going out of favor', which was clearly not the case. Cajal, in turn, referred to Golgi this way: 'We mourn this scientist who, in the last years of a life so well filled...' (Ramón y Cajal, 1906), in a manner not accurate for an active rector of a university and a senator, who would live another two decades. In doing so, it seems likely that the strong feelings engendered by their longtime rivalry led each of them to inadvertently slip into expressions of fantasy about the failure or mortality of the other.

The two antagonists were never to meet again. Cajal returned by train, passing up an appointment to receive yet another medal in Berlin, never again to visit the major scientific centers of Europe. The story is often told that upon arriving in Pavia, Golgi saw a band playing. While he reasonably enough thought that he was being honored for his achievement, it turned out to be a celebration for Giovanni Rossignoli, a celebrated bicyclist on his way back from his own victory, who by coincidence was also on the train.

REFERENCES for CHAPTER FOUR

Golgi, C.: The neuron doctrine – theory and facts. Nobel Lecture December 11, 1906.
https://www.nobelprize.org/uploads/2018/06/golgi-lecture.pdf

Ramón y Cajal, S.: The structure and connexions of neurons. Nobel Lecture December 12, 1906.
https://www.nobelprize.org/uploads/2018/06/cajal-lecture.pdf

CHAPTER FIVE: THE YEARS AFTER STOCKHOLM

Both men continued to have remarkably productive years after the awarding of the Nobel Prize, but it had been the climax of their rivalry, and the neuron doctrine had clearly won. As such, the events of the next years are less central to the theme of this book, which focuses on the dueling visions of Golgi and Cajal, and Golgi's unwillingness to change his mind in the face of contrary evidence. In order to provide context, however, we will briefly describe the lives of Golgi and Cajal after Stockholm, and touch on how the neuron doctrine has fared over time.

Golgi and Cajal

After returning home, Golgi's interest in actively engaging in research was rejuvenated. He resumed work from years before on cytoplasmic organelles in neurons, and by 1907 had modified the silver nitrate method further to visualize what came to be known as the Golgi apparatus. He was still active in politics, and in 1909 was engulfed in a controversy with his political rival Roberto Rampoldi, the Parliament member for Pavia, following the widespread distribution of a confidential letter Golgi had written to the mayor of Pavia criticizing Rampoldi. In the years before World War I, Golgi was also the chief of a medical division at San Mateo Hospital in Pavia, where he engaged in research on a variety of subjects including the healing of renal tissue and intestinal worm infections. During World War I he directed the Military Hospital in Pavia. He retired in 1918. He was honored with doctorates from a variety of institutions including the Sorbonne and the Universities of Athens, Geneva and Cambridge. In his last years he was very vocal in his resistance to the 1924 opening of the University of Milan, which he felt would detract from the fame of Pavia. He continued to publish research papers until 1923. He died at age 82 on January 21st, 1926.

A few months after returning from Stockholm, Cajal wrote a manifesto, 'The renaissance of the neuron doctrine', which among other things refuted the new variant on reticular theory resulting from the work on neurofibrils by Bethe and Apáthy (Chapter Three; Partsalis et al., 2013). He remained at the University of Madrid until he retired in 1922. In addition to his neuroanatomical work, he published on the histology of tumors, and like Golgi, studied infectious diseases including tuberculosis, syphilis, and cholera. He received many honors besides the Nobel Prize, Including becoming a member of the Royal Society of London, and receiving an honorary doctorate from Cambridge University. In 1920 he created the Spanish School of Histology in San Blas, which was later named after him. He continued to write as late as when he was on his deathbed, sending his former student Rafael Lorente de Nó (1902-1990) suggestions for staining techniques two days before the end (Ramón y Cajal, 1934). His final paper, published in 1934, the year he passed away on October 17 at age 82, was a detailed summation of neuron doctrine.

The evolution of the neuron doctrine

The neuron doctrine remains crucial to modern neuroscience, but in recent years, data have suggested that some of its aspects need to be reconsidered (Bullock, 2005). Although one of the foundational notions was that neurons are discrete anatomic units, in some cases they may not always be functional units (Nemiri, 2010). Since techniques for recording from multiple neurons at the same time became available, there has been growing interest in groups of neurons which emerge with their own functional properties (Yuste, 2015). Studies continue to suggest more complexity in how neurons are organized, so that the issue of individual neuronal action versus network functioning is still evolving (Shepherd, 2015).

The notion of neurons being discrete anatomical units has also been found to not be invariant. The mammalian nervous system includes neurons that have cytoplasmic continuity with their neighbors (Pereda, 2014), and which communicate by electrical means (Holzbecher and Kempter, 2022), leading them to fire synchronously (Bennett and Zukin, 2004; Pernelle et al., 2018). Electrical connections between neurons most commonly occur at gap junctions, which are comprised of two hemichannels, one on each of two neighboring cells. The channels formed by these junctions provide cytoplasmic continuity, and allow the passage of some ions including sodium and calcium and small molecules such as cyclic AMP, as well as electrical signals.

During early development of the mammalian nervous system, transient gap junctions rise in frequency and may be important for a number of processes including neuronal migration and differentiation, as well as the formation of synapses and circuits (Wang et al., 2012). Though found with lower frequency in most areas of the adult brain (Connors, 2009), their quantity can increase dramatically in response to injury such as ischemia and trauma. It appears that in various situations they may be involved in mechanisms of cell survival or death. It has been speculated, for instance, that in non-lethal injuries, gap junction signals may aid in survival, while in much more severe cases they may have the opposite effect (Belousov and Fontes, 2013).

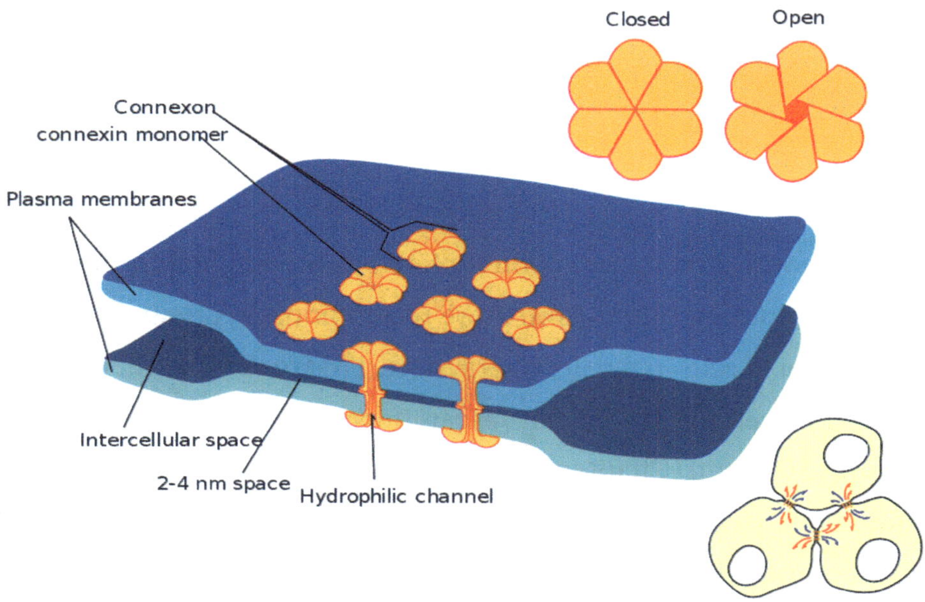

Figure 5-1: *Diagram of gap junctions connecting the plasma membranes (the large blue areas) of two cells. As described above, gap junctions are formed from two hemichannels, one in each plasma membrane, which connect to make a channel between the cells. Each hemichannel (connexon) is comprised of six individual proteins (connexins). On the upper right is an end-on view, showing the channel in closed and open positions.*

In a sense, then, some portions of the nervous system—particularly during early development—have proven to be different from the way Cajal's neuron doctrine envisioned it, although it should be noted that he himself suggested that 'neuronal discontinuity...could sustain some exceptions' (Swanson, 1995; Bullock et al., 2005). Insofar as gap junctions represent a kind of continuity between cells, they are reminiscent of Golgi's reticular theory. As it happens, the first electrophysiological and electron microscopic studies depicted discontinuity, with communication across chemical synapses. Some researchers (Shepherd, 2009) have wondered how early neuroscience might have developed if they had first detected electrical synapses at gap junctions.

Another consideration not contemplated in the original formulation of neuron doctrine stems from more recent findings about the role of non-neuronal cells in the nervous system. First recognized in the mid-nineteenth century by several researchers including Robert Remak, Theodor Schwann (Chapter One), and Rudolf Virchow (Chapter One), who coined the term *neuroglia* in 1856, they are numerically much more common than neurons by a ratio of three or more to one. Their name is derived from the Greek word *glia*, or glue, reflecting the original view that their role is largely structural.

As time went on, it became clear that although glia do not have true synaptic properties, they communicate with neurons by a variety of means involving ion fluxes as well as signaling and cell adhesion molecules, thus supporting the synaptic function of neurons and information processing (Fields and Stevens-Graham, 2002). Various types of glia may modulate the formation and maturation of synapses, offer metabolic support to neurons, and play a role in immune responses (Afridi et al., 2022). Although the roles of glia are not considered in the neuron doctrine *per se*, Cajal became interested in them, and went so far as to speculate that they might play a role in regulating sleep and wakefulness as well as higher functions including attention (Garcia-Mann et al., 2007). In a sense, then, he anticipated these much more recent advances.

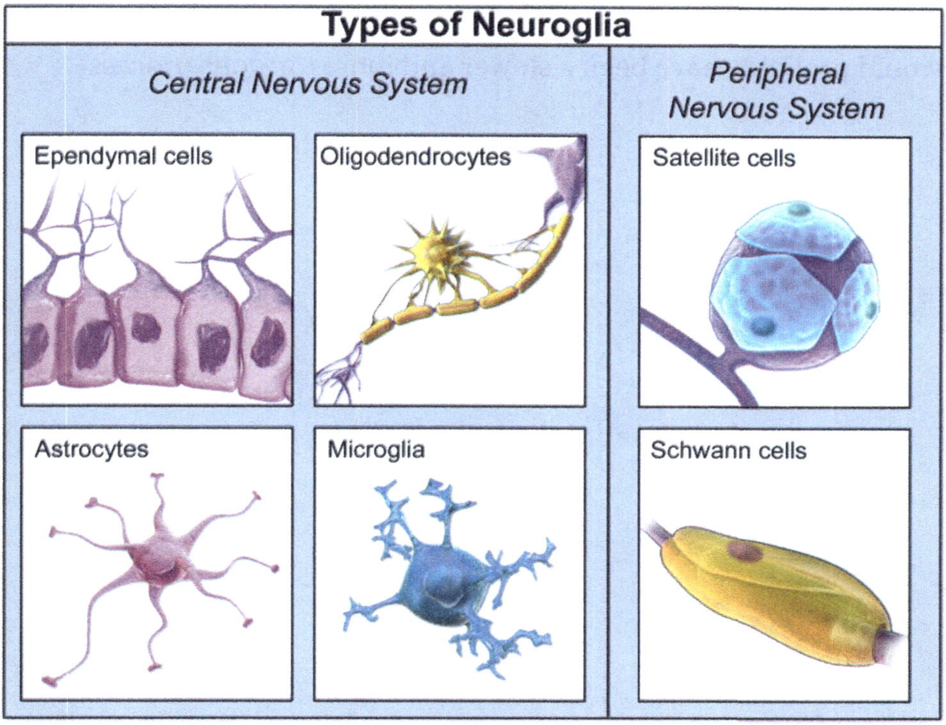

Figure 5-2: *Drawing of the major types of neuroglia. A detailed account of neuroglia and their function is found in von Bernhardi (2016).*

Stepping back to a broader view, though, the important point may not be whether the neuron doctrine, like most major biological models, has needed to be modified over the course of the next century. Rather, it is that its acceptance in the late nineteenth century led the way to our current understanding of the nervous system. Positing that neurons were discontinuous, for instance, set the stage for the next issue of how a neuron might excite or inhibit the subsequent one to which it is attached across a synaptic junction (Glickstein, 2006). It's likely we would ultimately have still come to the same understanding we have today without the debate about the neuron doctrine, but it would probably have been a slower and more tortuous process.

REFERENCES for CHAPTER FIVE

Afridi, R. et al.: The function of glial cells in the neuroinflammatory and neuroimmunological responses. Cells 11: 659, 2022.
https://www.mdpi.com/2073-4409/11/4/659/htm

Belousov, A.B.; Fontes, J.D.: Neuronal gap junctions: making and breaking connections during development and injury. Trends Neurosci. 36: 227-236, 2013.
https://www.ncbi.nlm.nih.gov/pmc/articles/PMC3609876/

Bennett, M.V.L.; Zukin, R.S.: Electrical coupling and neuronal synchronization in the mammalian brain. Neuron 41: 495-511, 2004.
https://www.cell.com/fulltext/S0896-6273(04)00043-1

Bullock, T.H. et al.: The neuron doctrine, redux. Science 310: 791-793, 2005.
https://www.science.org/doi/10.1126/science.1114394

Connors BW. Chapter 6: Electrical signaling with neuronal gap junctions. In: Harris A, Locke D, editors. Connexins: A Guide. Humana Press; 2009. pp. 143–164.

Fields, R.D.; Stevens-Graham, B.: New insights into neuron-glia communication. Science 298: 556-562, 2002.
https://pubmed.ncbi.nlm.nih.gov/12386325/

Garcia-Mann, V.; Garcia-López, P.; Freire, M.: Cajal's contributions to glia research. Trends. Neurosci. 30: 479-487, 2007.
https://pubmed.ncbi.nlm.nih.gov/17765327/

Glickstein, M.: Golgi and Cajal: the neuron doctrine and the 100th anniversary of the 1906 Nobel Prize. Current Biology 16: R147-R159, 2006.
https://www.cell.com/current-biology/pdf/S0960-9822(06)01203-6.pdf

Holzbecher, A.; Kempter, R.: Interneuronal gap junctions increase synchrony and robustness of hippocampal ripple oscillations. Eur. J. Neurosci. 48: 3446-3465, 2018.
https://onlinelibrary.wiley.com/doi/abs/10.1111/ejn.14267

Nemiri, A.: Santiago Ramón y Cajal. Scholarpedia 51: 8577, 2010.
http://www.scholarpedia.org/article/Santiago_Ramón_y_Cajal

Partsalis, A.M.; Blazquez, P.M.; Triarhou, L.C.: The renaissance of the neuron doctrine. Translational Neurosci. 4: 104-114, 2013.
http://www.leb.usp.br/PTC5736/partsalis_13.pdf

Pereda, A.E.: Electrical synapses and their functional interactions with chemical processes. Nat. Rev. Neurosci. 15: 250-263, 2014. https://www.ncbi.nlm.nih.gov/pmc/articles/PMC4091911/

Pernelle, G.; Nicola, W.; Clopath, C.: Gap junction plasticity as a mechanism to regulate network-wide plasticity. PLoS Comput. Biol. 14(3): e1006025, 2018. https://journals.plos.org/ploscompbiol/article?id=10.1371/journal.pcbi.1006025

Ramón y Cajal, S.: Letter to Rafael Lorente do Nó, October 15, 1934. Colegio Official de Médicos, Madrid, Spain.

Shepherd, G.M.: Neuron doctrine: historical background. In Encyclopedia of Neuroscience, 2009. Accessed September 6, 2022. https://www.sciencedirect.com/topics/veterinary-science-and-veterinary-medicine/reticular-theory

Shepherd, G.M.: *Foundations of the neuron doctrine: 25th anniversary edition, 2edn.* (New York, 2015; online edn., Oxford Academic. Nov. 1, 2015. https://academic.oup.com/book/24483

Swanson, N and Swanson, L.W.l.: *Histology of the nervous system of man and vertebrates by S. Ramón y Cajal.* Oxford University Press, New York, 1995 edition.

https://scholar.google.com/scholar_lookup?title=Histology+of+the+Nervous+System+of+Man+and+Vertebrates%2C&author=S.+Ramòn+y+Cajal&author=N.+Swanson&author=L.W.+Swanson&publication_year=1995

Shepherd, G.M.: Neuron doctrine: historical background. In: *Encyclopedia of Neuroscience* (L.R. Squire, ed.), Academic Press, 2009.
https://www.sciencedirect.com/topics/veterinary-science-and-veterinary-medicine/reticular-theory

von Bernhardi, R.: *Glial cells in health and disease of the CNS.* Springer, 2016.
https://link.springer.com/book/10.1007/978-3-319-40764-7

Wang, Y. et al.: Neuronal gap junction coupling is regulated by glutamate and plays critical role in cell death during neuronal injury. J. Neurosci. 11: 713-725, 2012.
https://www.ncbi.nlm.nih.gov/pmc/articles/PMC3567463/

Yuste, R.: From the neuron doctrine to neural networks. Nature Reviews Neuroscience. 16: 487-497, 2015.
https://www.nature.com/articles/nrn3962

CHAPTER SIX: SUMMARY

Golgi and Cajal did not directly interact until around 1889, when they were 46 and 37 years old respectively. In Chapters Two and Three we described their backgrounds and formative years separately. In closing it may be worthwhile to juxtapose them, in order to see their similarities and differences, and how their formative years and temperaments may have influenced how they dealt with one another.

In a sense, Golgi and Cajal came from similar backgrounds in small villages in which their fathers were the local physicians. Their fathers, the sons of a hat maker and an impoverished farmer respectively, had seen medicine as a way of social advancement. Their temperaments and relations with their own sons were very different, however. Golgi's father, who as a member of a mercantile family had a somewhat simpler process, had entered The University of Pavia's medical school, graduated in 1838, and began a career in which he was a respected member of the community; it seemed a natural, and uncontested, decision that the son in turn would go into medicine. In contrast, Cajal's father, from a destitute family, had to struggle step by step from becoming a barber's apprentice to finally getting a medical degree, all the while supporting his family. The process had involved frequent moves, either to better educational opportunities or, more commonly, to successive villages after disputes about his compensation. He felt passionately that his son, too, should become a doctor, and was infuriated by his reluctance; their arguments about his future plans were often punctuated by beatings.

The temperaments of the two sons were also very different. Golgi was a quiet, diligent student, and aside from one episode in which he was caught up in the local resentment of the Austrian occupation, did well in secondary school. He had always planned to go to medical school, and initially thought he would practice clinical medicine. In contrast, Cajal was a rebellious youngster whose pranks even led him to a few days in jail. He loved drawing, and became enamored of the romance and adventure he discovered in books filched during his escapades. It was only after much resistance that he was ultimately persuaded by his father to become a doctor.

Once in medical school, the two had very different experiences. Golgi was inspired by two teachers, Cesare Lombroso and Giulio Bizzozero, and became enthusiastically involved in Bizzozero's pathology laboratory. Upon graduation, Golgi took the only position available to him as a clinical doctor, a role in which he felt comfortable, but his heart was in the research which he pursued as much as he could. In contrast, Cajal was largely unmoved by his time in medical school, skipped many classes, and after graduating in 1873, was not eager to practice medicine. After recovering from his misadventure in Cuba, he briefly assisted his father in his practice; it became clear that he was not well suited for clinical medicine, and he took a job teaching anatomy in a provincial medical school. In wasn't until 1877, when at the insistence of his father he traveled to Madrid to take his doctoral exams, that he found his calling: a visit to the laboratory of Aureliano Maestre de San Juan introduced him to the fascination of microscopy. It is possible that this appealed to his lifetime interest in exploration, and saw himself as discovering new worlds. In any event, he quickly bought an instrument and set up a laboratory in his father's home.

For the two, then, involvement in microscopy had different meanings: for Golgi, it was a natural evolution in following the guidance of a respected teacher. For Cajal, who was then drifting in some senses, it was the first time he found something professionally that caught his imagination. Once involved, however, both men had similar reactions, throwing themselves into their work with single-minded persistence. Fortunately, setting up a histology laboratory was relatively inexpensive, and the nature of the work suited them both very well: it was an activity that could be conducted by a single person. In some senses the original settings of their work were not so different, with Golgi alone in the monk-like atmosphere of his kitchen laboratory, and Cajal in his father's attic, and later in rooms connected to his apartment. The intensity of their work led to relatively rapid results as well: Golgi began at the chronic hospital in Abbiategrasso and built his laboratory starting in June of 1872, and published the first results of his method in August,1873; Cajal, returning from Madrid armed with his new-found knowledge of Golgi's method in January 1887, published his first account of the anatomy of individual neurons in May, 1888.

In addition to their resolute and tenacious manner, another similarity between the two was of course their brilliance. It has never been absolutely clear how Golgi thought of the black reaction. There was not an obvious theoretical foundation that might have led him to it, and both potassium dichromate and silver nitrate had been individually used in histology without promising results. Somehow, he recognized that the two could be used sequentially, leading to his discovery. It was a difficult technique, however, with inconsistent results, and was not taken seriously by histologists over the course of the next 14 years. It was Cajal's brilliance to see the potential in it that others had overlooked, and then to set about refining it.

Both men also had to struggle with a similar handicap: neither Italy nor Spain in those years were considered to be in the forefront of European science, which was conducted in Berlin, Vienna, Paris, and London. Perhaps not surprisingly, the prominent scientists in these glamorous capitals tended to be skeptical of claims of breakthroughs by researchers from areas they thought to be less central, whose papers were in languages they often did not read. Golgi's first publication in 1873 about the black reaction was in the relatively obscure *Gazzetta Medica Italiana*; Cajal self-published his description of individual neurons in 1888, printing only 60 copies of his *Trimonthly Review of Normal and Pathological Histology,* which he mailed to other scientists with essentially no responses. Cajal's breakthrough moment was when he attended an international meeting in Berlin in 1889 and persuaded Albert Kölliker of the importance of his work. Subsequently, Golgi was distressed to see how his name, with the discovery of the black reaction, was initially often mentioned in international circles in conjunction with Cajal's.

These two scientists, who shared a passion for their work and the struggle to be recognized in the mainstream of European science, were also separated, by temperament and timing of their careers. Cajal, whose aggression and anti-authoritarian tendencies in his youth had often required him to change schools, even to be briefly jailed, was not the gentlest of antagonists. When writing Golgi on August 27, 1889, to suggest that he visit to discuss their differences, he characterized Golgi's position as 'absurd' (Ramón y Cajal, 1900; Ehrlich, 2022), perhaps not the optimal technique for establishing a dialogue. As described in Chapter Four, when Golgi arrived in Stockholm in 1906, Cajal was there with the party to greet him at the train station, but Golgi was very abrupt, leaving for his hotel instead of talking. This has usually been interpreted as Golgi passing up a last opportunity for mutual understanding, but it seems equally likely that Cajal came for confrontation, not reconciliation. While Cajal was very comfortable being in the vanguard of a new movement, Golgi was unafraid to be seen as supporting an older viewpoint. In his Nobel lecture, he noted that 'I cannot abandon the idea of a unitary action of the nervous system, without bothering if, by that, I approach old conceptions' (Golgi, 1906). He was, then, in the words of his biographer, a 'conservative revolutionary' (Mazzarello, 2019), very different in spirit than Cajal.

One can only speculate about whether qualities of Golgi's personality contributed to his struggle against the neuron doctrine. It may be that he was a person who tended to double-down and dig in when challenged, though your author has not found clear information one way or the other in his biographies. It is possible, though, to infer something about his conservatism and reluctance to change from one other piece of his history. As described in Chapter Five, he spent his last years in an unsuccessful struggle to prevent the creation of a new university in Milan, apparently because he feared it would take away from the glory of his own nearby University of Pavia. Thus, his efforts to maintain the reticular theory in the face of new developments was not the only case in which he fought on, and lost, in an effort to maintain the prestige of something to which he was attached.

Despite their differences, Golgi and Cajal's passion for histology also led to a commonality, their seeming lack of interest in the growing contributions of electrophysiology to understanding the relations between neurons. Cajal, focusing on individual cells, did not seem as intrigued by what went on in the space between them. Charles Scott Sherrington, who had coined the term synapse, found room for Cajal's work on the individual nature of neurons, as the need for signals to cross the synapse helped explain the time lag he had demonstrated in spinal reflexes in the 1890s, but Cajal did not seem to return the favor. It was not due to being unaware of Sherrington's work: Sherrington had visited Cajal in Spain in 1885 (which contributed to Sherrington's interest in studying the spinal cord), and Cajal had stayed with him in London for two weeks in March 1894 when he was invited to give a talk to the Royal Society. As mentioned earlier, Sherrington went on to challenge Cajal's concept of dynamic polarization, or one-way movement of impulses along neurons, with his finding that excitation could move in both directions inside a single neuron, though not in a series of them because this was prevented by the gate-like function of synapses, a word Cajal very rarely used.

It has been suggested that both Golgi and Cajal tended to overlook the new discoveries of neurophysiology because of their belief that the nervous system was best understood though microanatomy (Berlucci, 1999). It would have been interesting to see how they might have reacted to Edgar Adrian's (1889-1997) demonstration of the neuronal action potential in 1925, shortly before Golgi's death and in Cajal's later years. Part of their lack of interest may have been a matter of timing, as Sherrington's electrophysiologic findings were appearing after their differing, but firm, positions on the nature of neurons were well established. But the broader point is that in retrospect their mutual passion for histology, which led to their remarkable discoveries, may also have ironically limited some aspects of their vision for the future. Cajal's youngest student, Lorente de Nó (1902-1990), tried to gently suggest this to Cajal, writing from America, for instance, that in his work 'the mornings are dedicated to anatomy and the evenings to physiology' (Lorente de Nó, 1933). It lay with him and others of the next generation to help guide Cajal's school from a focus on anatomy to include neurophysiology (Espinoza-Sanchez, 2019).

The timing of Golgi's and Cajal's careers may also have played a role in their own interactions. In 1889, when Cajal wrote Golgi attacking his position, he was a new chairman of histology in Barcelona, still on the rise in his field and only to achieve the pinnacle of Spanish academe—a chairmanship in Madrid—for another three years. In contrast, Golgi was the chairman of general pathology in Pavia, and a powerful administrator who would become rector of the university in 1893. As such, it was likely that he was not used to being told that his ideas were 'absurd', especially by a Spanish researcher who was not well known outside his country. Unlike Cajal, whose driving passion was neurohistology, Golgi had moved on. He had primarily been working on infectious diseases and other topics since 1885 and had not been personally engaged in histology of the nervous system for several years (Ehrlich, 2022, p. 179). His response to receiving Cajal's paper in 1889 had been to mail him a copy of his malaria work. If instead he had been actively devoted to studying brain histology at the time, one wonders if his response might have been different, and even whether he might have been more open to considering a newer theory of neuronal function.

REFERENCES for CHAPTER SIX

Berlucchi, G.: Some aspects of the history of the law of dynamic polarization of the neuron. From William James to Sherrington, from Cajal and van Gehuchten to Golgi. J. Hist. Neurosci. 8: 191-201, 1999. DOI: 10.1076/jhin.8.2.191.1844

Ehrlich, B.: *The Brain in Search of Itself: Santiago Ramón y Cajal and the Story of the Neuron*. Farrar, Straus and Giroux, New York, 2022.

Espinosa-Sanchez, J.M.; Espinosa-Campos, L.; Batuecas-Caletrio, A.: Lorente de Nó: From neuroanatomy to neurophysiology. The Anatomical Record 303:1221-1231, 2019. https://anatomypubs.onlinelibrary.wiley.com/doi/full/10.1002/ar.24190

Golgi, C.: The neuron doctrine – theory and facts. Nobel Lecture December 11, 1906. https://www.nobelprize.org/uploads/2018/06/golgi-lecture.pdf

Lorente do Nó, L.: Letter to Santiago Ramón y Cajal, May 28, 1933. Archivo Fernando de Castro, Censo-Gula de Archivos de Espana e ibeoramérica #ES.28079.AFC; Madrid, Spain.

Mazzarello, P.: Camillo Golgi: the conservative revolutionary. Italian Journal of anatomy and embryology. 124: 288-304, 2019. https://oajournals.fupress.net/index.php/ijae/article/download/11658/11088/

Ramón y Cajal, S:: *La Correspondencia Médica.*August 18, 1900.

PICTURE CREDITS

The author's assessment is that all images are in the public domain or presented under the terms of Section 107 of the U.S. Copyright Law (the 'Fair Use' provision). When appropriate, all reasonable efforts have been employed to trace copyright holders and to get their permission for the use of copyright material. The author apologizes for any errors or omissions in this list and will gratefully include any corrections in future editions if notified.

Cover illustration: Modified image of histological sample of nervous system tissue by Ramón y Cajal, who referred to some neurons as 'butterflies of the soul'. Original supplied by the Cajal Institute (CSIC), 1900, from Wikimedia Commons, which states 'This work is in the public domain in the United States because it was published (or registered with the U.S. Copyright Office) before January 1, 1927. '

Figure 1-1: (Nimrud lens) Photo of the Nimrud lens in the British Museum by Geni, in Wikimedia Commons, which states 'This file is licensed under the Creative Commons Attribution-Share Alike 4.0 International, 3.0 Unported, 2.5 Generic, 2.0 Generic, and 1.0 Generic License.

Figure 1-2: (Hooke's microscope) Hooke, R., image mounted by Alejandro Porto, in Wikimedia Commons, which states 'The author died in 1702, so this work is in the public domain in its country of origin and other countries and areas where the copyright term is the author's life plus 100 years or fewer.'

Figure 1-3: (Miracle of the slave). Tintoretto painting, 1548, photo source unknown, from Wikimedia Commons, which states 'This work is in the public domain in its country of origin and other countries and areas where the copyright term is the author's life plus 100 years or fewer.'

Figure 1-4: (Logwood) Adolphus Ypey, 1813, in Wikimedia Commons, which states 'This work is in the public domain in its country of origin and other countries and areas where the copyright term is the author's life plus 70 years or fewer.'

Figure 1-5: (Spinal cord motor neurons) Yurvik55 in Wikimedia Commons, which states 'This file is licensed under the Creative Commons Attribution 4.0 International License.'

Figure 1-6: (camera obscura) Unknown illustrator, 1850, in Wikimedia Commons, which states 'This work is in the public domain in its country of origin and other countries and areas where the copyright term is the author's life plus 70 years or fewer.'

Figure 1-7: (heliograph of a man leading a horse) Nicéphore Niépce, 1825, from Wikimedia Commons, which states 'This work is in the public domain in its country of origin and other countries and areas where the copyright term is the author's life plus 100 years or fewer.'

Figure 1-8: (diorama) Drawing of diorama , modified from sketch in Almanach des Spectacles, 1823, p. 240, found in Wikimedia Commons, which states 'This work is in the public domain in its country of origin and other countries and areas where the copyright term is the author's life plus 70 years or fewer.'

Figure 1-9: (Boulevard du Temple) Modified from picture by Louis Daguerre, from Wikimedia Commons, which states 'This work is in the public domain in its country of origin and other countries and areas where the copyright term is the author's life plus 100 years or fewer.'

Figure 2-1: (University of Pavia) Unipavia, 2021, from Wikimedia Commons, which states 'This file is licensed under the Creative Commons Attribution 2.0 Generic License.'

Figure 2-2: (Golgi drawing) Camillo Golgi, 1875; J. DeFeloipe in Wikimedia Commons, which states 'This file is licensed under the creative Commons Attribution-Share Alike 4.0 International License.'

Figure 2-3: (Golgi) Unknown author, 1903. From Wikimedia Commons, which states 'This work is in the public domain in its country of origin and other countries and areas where the copyright term is the author's life plus 70 years or fewer.'

Figure 3-1: (Petilla) Modified from image by JLPedrafita in Wikimedia Commons, which states 'This file is licensed under the creative Commons Attribution-Share Alike 4.0 International License.'

Figure 3-2: (Cajal 1884-1887 in Valencia) Santiago Ramón y Cajal, from Wikipedia Commons, which states 'The author died in 1934, so this work is in the public domain in its country of origin and other countries and areas where the copyright term is the author's life plus 80 years or fewer.'

Figure 3-3: (Simarro) Legado Luis Simarro, Fundacion General in Wikimedia Commons, which states 'This work has been released into the public domain by its author.'

Figure 3-4: (dendritic tree) Michael Hausser, in the Wellcome Collection, which states that this file is licensed under the 'Attribution-NonCommercial 4.0' license.

Figure 3-5: (Cajal, 1899). Anonymous, restored by Garrondo, in Wikimedia Commons, which states 'This work is in the public domain in the United States because it was published (or registered with the U.S. Copyright Office) before January 1, 1927.'

Figure 3-6: (synapse) BruceBlaus, in Wikimedia Commons, which states 'This work is free and may be used by anyone for any purpose.'

Figure 5-1: (gap junction) Mariana Ruiz LadyofHats, in Wikimedia Commons, which states 'This work has been released into the public domain by its author, LadyofHats. This applies worldwide.'

Figure 5-2: (neuroglia) BruceBlaus, in The Medical Gallery of Blausen Medical 2014, from Wikimedia Commons, which states 'This work is free and may be used by anyone for any purpose.'

Made in the USA
Coppell, TX
18 January 2026

69483036R00066